Dictionary of
Energy and Fuels

Clifford Jones

Reader in Engineering, University of Aberdeen, UK and

Nigel Russell

Lecturer in Chemical Engineering, The University of Sheffield,
UK

Whittles Publishing

CRC CRC Press
Taylor & Francis Group

Published by
Whittles Publishing Limited,
Dunbeath,
Caithness, KW6 6EY,
Scotland, UK
www.whittlespublishing.com

Distributed in North America by
CRC Press LLC,
Taylor and Francis Group,
6000 Broken Sound Parkway NW, Suite 300,
Boca Raton, FL 33487, USA

ISBN-13   978-1-904445-44-9
USA ISBN   978-1-4200-6194-9

Printed by Athenaeum Press Ltd., Gateshead, UK

Contents

Dedicated to the authors' mutual friend
Jack Nicholls
Bishop of Sheffield

Introduction

This book contains almost 1500 entries, presented alphabetically, each relating to fuels, energy and combustion. A few entries are concerned with basic principles of combustion, and all solid, liquid and gaseous fuels have been covered across the entries. There are also entries concerned with nuclear energy, mainly in relation to electricity generation. Other entries are concerned with supply and demand of oil and gas, there being information for example on oil tankers, pipelines and related economics. There are a number of linguistic entries and a few which are sociological, political or historical. There has been an attempt to blend major topics which will have a long shelf life and be useable to a reader thirty years down the track with topics of particular interest to the fuel technologist in the early 21st century. Comprehensiveness in the sense of completeness cannot possibly be achieved in such a volume. However, comprehensiveness in a different sense can be aimed for by having a skilfully balanced variety of entries.

Who then will benefit from such a book? We hope that students in such disciplines as chemical engineering and petroleum engineering will in fact benefit in these ways. First of all they can go to the book for information whereby they can follow up an idea implanted during a lecture or tutorial. Secondly, frequent reference to the book will build up in a student's mind a sense of how major topics studied fit into the past, present and future world energy scene. The word 'world' is used advisedly as the entries have been written so as to inform a reader of what is taking place where in terms of fuel production and utilisation. So whilst a reader will find information herein on the Gulf of Mexico and on Saudi Arabia, he or she will not look in vain for information on a 'poor' country like Malawi. Selected entries from the book can be used by a lecturer as a teaching resource with a number of interlinked entries readily forming a discussion point for a tutorial.

Finally, we hope that the book might serve as a reference, for example in organisations where not all staff are trained in the subject area and not all undertakings are fuels related. It is doubtful whether a complete 'lexicon' of fuel technology could or should be attempted, but this book will serve that purpose to a limited degree. The cross references (marked in bold) will assist a reader so using the book in finding related entries or in locating the entry closest to that being sought. We also venture to hope that there will be some readers already knowledgeable in the subject area who will find that a few new insights into certain issues have been given and that the book is therefore of interest.

The sources – some conventional but most of them electronic – upon which we have drawn are far too numerous to cite individually. We can only hope that the requirements both of courtesy and of publishing practice are met by our making a single acknowledgement of gratitude to the originators of all of the sources we have used.

Energy supply and demand is going to continue to drive international affairs. The world needs men and women trained as specialists in fuels and energy. It also requires men and women trained in contiguous disciplines who have a broad insight into energy matters. It *also* needs men and women with no professional involvement at all in energy matters who nevertheless take an intelligent interest. This book is directed at all three groups.

JCJ & NVR

Abadan Refinery
Anglo-Persian Oil Company in **Iran**, capacity 180 000 **bbl** per day.

Abqaiq
Scene of a **giant field** in **Saudi Arabia**, second only to the **Ghawar** field, with proven reserves of 17 billion **bbl** of oil. Abqaiq is also the scene of a major facility which prepares crude oil from all Saudi sources for export. The crudes of Saudi Arabia vary in density and 'sweetness' and can be adjusted in quality by blending before export. There is also likely to be **desulphurisation**.

Acacia power station
Small power station near Cape Town owned and operated by **Eskom**. The Acacia facility came into operation 30 years ago and uses three gas turbines. Electricity production is up to 171MW. In February 2006 proposals for a merger between Eskom and **Sasol** were rejected by the South African Competition Tribunal.

Acetone
Low flash-point organic liquid, the simplest ketone, having structural formula:

$$H_3C-\overset{\displaystyle O}{\underset{\displaystyle \|}{C}}-CH_3$$

There have been many trials of acetone as a fuel additive both for spark-ignition and from compression-ignition engines. In one of the most recent (April 2005) a 6–7% reduction in diesel consumption was reported for a heavy semi-trailer in long-distance use in the US when two ounces of acetone were added to each 10 gallons of fuel supplied to the vehicle. The test was over 12 450 miles of travel. Similar fuel savings and also performance improvement have been observed with acetone as an additive for gasoline. Acetone in industrial quantities can be made from **synthesis gas** followed by **oxidative coupling** or by the partial oxidation of propane.

Acetylene
Organic compound, structural formula HC≡CH, widely used as a fuel in a premixed flame with oxygen in applications such as welding and flame cutting. It can be manufactured by the dehydrogenation of **ethylene** obtained from petroleum sources.

Acid rain
A popular term for the deposition of acidified precipitation. Oxides of nitrogen and sulphur originating from the combustion of fossil fuels undergo complex chemical reactions in the atmosphere. They eventually form sulphuric and nitric acids, subsequently precipitating hundreds, or even thousands, of miles from the original pollutant source. Acid rain has affected forests, lakes and buildings. Fossil fuel emission reduction technology has reduced NO_x and SO_2 emissions. However, the increase of **fossil fuel** combustion globally has been matched by an increase in acid rain.

Acusorb
Proprietary substance which, when prepared in the form of small beads, is used to remove water from **biodiesel** fuels. Having once reached their maximum water holding capacity they can be regenerated by heating. Two hundred such regenerations are possible before the material is regarded as 'spent'.

Adiabatic flame temperature
The temperature of the products in a combustion process that occurs without any heat transfer, work, or changes in kinetic or potential energy. This is the maximum temperature of the products because any heat transfer from the reactants would lower the temperature of the products.

AFFF
See **Aqueous film-forming foams**

Afghanistan, energy scene in

This country is self-sufficient in natural gas, at present importing and exporting none. It is however totally reliant on exports of crude oil amounting to about 3500 **bbl** per day. There is coal, but its current production is as low as 1000 **short ton** per year, having been two orders of magnitude higher in the early 1990s.

The capital Kabul receives hydroelectric power from three dams and there is thermal electricity generation using natural gas as fuel at about 45 MW. There is also some import of electrical power. At the present time, only about 10% of the population of Afghanistan has access to electricity.

Africa, hydroelectric power in

The **Kariba dam** is but one example (though certainly a noteworthy one) of **hydroelectricity** in the continent of Africa; it provides power for Zimbabwe and for Zambia with some left over for export. South Africa has a hydroelectric generating capacity of 650 MW (0.65 GW). In fact, 45 African nations, of widely varying social and political milieu, are in the business of **hydroelectricity** at the present time. The 2005 operating capacity summed for the 45 countries is in excess of 20 GW and it is certain that there is much scope for expansion. The African country with the largest capacity for hydroelectric power generation is Egypt because of its famous Aswan dam which can produce power at almost 3 GW. Mozambique raises power hydroelectrically at 2.2 GW. Others with high levels of power so generated include Morocco (1.2 GW), Ethiopia (0.8 GW), Kenya (0.6 GW) and the Ivory Coast (0.3 W). Of course, the hydroelectric generation will not work to full capacity all the time. Kenya is examined in the example below, where six hydroelectric installations with capacities 40–144 MW provide a total of 0.6 GW. Capacity and actual generation data are for 1999, the most recent year for which information is available.

Generating capacity 0.6 GW

Quantity of electricity made by this means in 1999 = 3294 GWh

Percentage of full capacity usage =

$$\frac{3294 \text{ GWh}}{0.6 \text{ GW} \times (24 \times 365)\text{h}} \times 100$$

$$= 63\%$$

The landlocked country of Lesotho has a hydroelectric capacity of 79 MW. This is very significant indeed in a nation where fewer than 3% of the population have electricity supplied to their homes. A similar calculation for Lesotho reveals that hydroelectric generation is apparently only used to 30% capacity. This is no doubt due to the fact that to connect one building to the national power supply costs about $US 1000, a gigantic sum of money for a typical resident of Lesotho. As noted in subsequent entries there is hydroelectricity in Burundi and also in Malawi.

Africa, role of in world oil growth

Nigeria, Libya and Algeria (all **OPEC** countries) are major producers of oil. In southern Africa, Angola is the only major producer although*, as described in subsequent entries, Namibia is becoming a significant producer and South Africa is hopeful of the discovery of oil off the Kwazulu-Natal coast. Highly significant is the fact that a third of the new oil discoveries in the first five years of the 21st Century have been in Africa.

Afterburning

Incineration of the effluent gas in order to break down toxic or odorous substances. If the gas contains an appreciable amount of **carbon monoxide**, as is sometimes the case, it may be ignitable as it is. Otherwise, blending with natural gas is necessary. Catalytic afterburning is possible, in which case temperatures can be as low as 300°C.

Afterdamp

In mining, the gas resulting from the explosion of **firedamp**, containing **carbon monoxide**. Also known as 'chokedamp'.

Air-fuel ratio

An expression of the relative amounts of air and fuel in a combustion system by either mass or volume.

Aircraft, carbon dioxide emissions from

Sub-sonic jet aircraft fly at heights within the troposphere, the lowest

* Note added in proof: Angola became a member of OPEC on 1st March 2007.

layer of the atmosphere. Commercial aircraft of the world release about 600 Mt of carbon dioxide annually, about 7% of the total release.

Airline fuel surcharge

An airline might need to impose a fuel surcharge on top of the basic cost of a particular ticket if oil prices increase unexpectedly. British Airways imposed such a surcharge in June 2005 when the cost of a **barrel** of crude oil approached $US 60. The Australian airline Qantas have also been imposing such surcharges as have, amongst others, the Dallas/Fort Worth based South West Airlines, a budget domestic carrier, though their surcharges have been lower than those of other US airlines operating the same routes.

Airports, supply of fuel to

Fuel at an airport will be stored for use at a 'fuel farm' comprising a number of tanks. For example, the fuel farm at Newcastle Airport, UK, one of the most rapidly expanding airports in Europe at the present time, has three tanks each of 680 000 litres for jet fuel and one 44 000 litre tank for **avgas**. Each tank is surrounded by a bund which will prevent spread in the event of tank failure. Management and operation of the fuel farm are contracted by the British Airports Authority to Swissport International. Fuel is received from four companies who were successful in their bid to supply the airport: Texaco, Conoco-Phillips, BP and **Q8**. A quantity of 100 Mt of fuel per year passes from the fuel farm to aircraft.

At airports around the world, pipeline delivery to the fuel farm is the norm, e.g. the Fina-Line pipeline which connects an oil refinery in East Anglia to Heathrow Airport. This pipeline is 140 miles in length and is capable of delivering 6 million litres of fuel per day.

Akkuyu

Scene of two proposed nuclear power plants for Turkey. The smaller plant, estimated to be completed within the next few years, will be capable of generating 650 MW of electricity. The second, not expected to begin producing until after 2010, will have a capacity twice this. At present, Turkey has no operating nuclear power stations.

Alaska, North Slope of

A scene of oil and gas production. The total estimated quantity of oil is 34 billion **bbl**, less than half of which has been extracted during 30 years of operations.

Alba field

One of the more recent oil fields in the North Sea to have been developed, the Alba field was discovered by Chevron in 1984; oil production began in 1994. The field contains a billion **bbl** of oil. The wells are in shallow water but deep drilling—about 3000 m—into the seabed was required hence the product of the field is classed as deep sea oil. **Deep-sea oil and gas** is largely associated with the Gulf of Mexico and Chevron have claimed with some justification that Alba belongs to a 'new generation' of North Sea field. The platform originally installed primarily for drilling was fixed, not floating, and later became part of the production platform. Facilities include a floating storage unit and a 10 MW gas turbine electricity generator: the former enables the oil to be transferred to tankers rather than taken to an onshore terminal by pipeline. The output from the field is 100,000 bbl per day.

Also a condensate field in waters off Equatorial Guinea, estimated contents 140 billion m^3 of gas. At present 7 million m^3 of gas per day are produced at the field which on stripping yields 19 500 bbl per day of liquid. Some of the gas after stripping is converted to **methanol** at a plant on Bioko Island in which Marathon have a 45% holding. The methanol is exported to Europe and the USA.

Albania, energy scene in

This country has strong reserves of coal, oil, gas, and wood as well as some **hydroelectricity**. The general malaise of the region has however precluded benefits from these assets. Oil production is a mere 3 million **bbl** per year, a fifth of what it was 30 years ago. Oil and gas production have been adversely affected by failure to replace aging equipment and general neglect. The story with regard to coal is similarly dismal. As a means of improving matters it is proposed that Albania shall be linked by a pipeline to one or more of the more advanced European countries ensuring, between the domestic and imported quantities, a reliable supply. Licensee companies are exploring the Adriatic for offshore hydrocarbon reserves. Foreign help is also being obtained in enhancing the productivity of the current onshore fields.

The country's electricity generation, 11 650 MW, is 87% hydro and 13% thermal.

Alberta, plans to increase the refining capacity of

Arizona Clean Fuels Yuma refinery is the first **grass roots refinery** project in the US for 30 years. Nevertheless, increased refining capacity for N. America will be required as the 21st century enters its second decade.

Indications are that Alberta, Canada, rather than a US location will be the scene of a new 450 000 **bbl** per day refinery.

Alkaline fuel cell

Alkaline fuel cells use an electrolyte that conducts hydroxyl (OH^-) ions from the cathode to the anode. The electrolyte is typically composed of a molten alkaline mixture such as potassium hydroxide (KOH). They operate at 65–220°C and a pressure of about 15 psig (1 barg). Each cell can produce up to between 1.1 and 1.2 V_{DC}.

The advantages of alkaline fuel cells are that they:
- operate at low temperature
- have fast start-up times (50% rated power at ambient temperature)
- have high efficiency
- need little or no expensive platinum catalyst
- have minimal corrosion
- have relative ease of operation
- have low weight and volume

The disadvantages are that they:
- are extremely intolerant to CO_2 (about 350 ppm maximum) and somewhat intolerant of CO. This means the oxidant must be either pure oxygen or air that has been scrubbed free of carbon dioxide. The fuel must be pure hydrogen due to the presence of carbon oxides in reformate
- have a liquid electrolyte, introducing liquid handling problems
- require complex water management
- have a relatively short lifetime

Alkaline fuel cells must operate using pure hydrogen free of carbon oxides. The reactions at the anode are:

$$H_2 + 2K^+ + 2OH^- \rightarrow 2K + 2H_2O$$

$$2K \rightarrow 2K^+ + 2e^-$$

The reactions at the cathode are:

$$\tfrac{1}{2}O_2 + H_2O \rightarrow 2OH$$

$$2OH + 2e^- \rightarrow 2OH^-$$

The OH^- ion is drawn through the electrolyte from the cathode to the anode by the reactive attraction of hydrogen to oxygen, while electrons are forced through an external circuit from the anode to the cathode. Combining the anode and cathode reactions, the overall cell reactions are:

$$H_2 + 2OH^- \rightarrow 2H_2O + 2e^-$$

$$\tfrac{1}{2}O_2 + H_2O + 2e^- \rightarrow 2OH^-$$

The fuel cell produces water that either evaporates into the source hydrogen stream (in an immobile system) or is flushed out of the cells along with the electrolyte (in a mobile system). This water must be continually removed to facilitate further reaction.

Alpha olefins

Linear (i.e. not branched) **olefins** made from **ethylene**, being very versatile in petrochemical manufacture. They are of primary interest in fuel technology, being the base of **synthetic motor oils**.

Alqueva project

Construction whereby hydroelectricity, irrigation and water will be supplied in an integrated way by means of two dams in Portugal about 70 miles from Lisbon. Completed in 2002, it provides several hundreds of MW of power adding to the already very significant **hydroelectricity** in Portugal.

Altamont Pass, CA

Scene of a very large wind farm which produces over 500 MW. It is one of the oldest wind farms in the US and one of the largest concentration of wind turbines in the world. However, an ecological effect of an operation is usually in proportion to its magnitude, and at Altamont the problem of birds being killed by **wind turbines** is acute; a lawsuit is pending. About a thousand birds including eagles, owls and American kestrels are killed each year at this facility which is on a migration route taken by the birds.

Alumina

Oxide of aluminium, chemical formula Al_2O_3. It occurs as a constituent of the clay content of coals.

Aluminium production, power requirements of

A third of the cost of production of aluminium from bauxite—an electrolytic process—is accounted for by the electricity required. Canada is a major aluminium producer, where it is estimated that 15 kWh of electricity are used per kg of aluminium. While in Canada hydroelectric power is widely used in the aluminium industry, power is generated by thermal stations in other countries such as Australia. Because electricity accounts for such a major component of the cost of aluminium, it sometimes happens that in aluminium production electricity charges are indexed to profits. This means that the electricity supplier may accept a lower price per unit of electricity if sales of aluminium are down.

Alve field
Condensate field in the Norwegian sector of the North Sea, expected to be producing by late 2008 making use of infrastructure at an adjacent field. Estimated reserves are 6.8 billion m^3 of gas and 8.3 million **bbl** of condensate.

Ambient standard
The maximum level imposed by law of a particular gaseous pollutant in the atmosphere, distant from any one powerful releaser such as a power station. In going from the **emission standard** which applies at a major releaser to the ambient standard there is dilution by about a factor of 10^3. Ambient standards for **sulphur dioxide** are of the order of parts per hundred million.

Anaran field
Oil field in western Iran, currently being developed. The Norwegian company Norsk Hydro and **Lukoil** are both participants. The field is expected to produce 10 000 **bbl** per day of crude oil by 2010. Iran has a buyback system for foreign enterprises operating its oilfields and 30% of the oil from Anaran will be retained by Iran under such an arrangement.

Andhra Pradesh
Scene of an newly discovered gas field off the southwest coast of India, believed to contain ≈500 **bcm** of natural gas. Development will be the responsibility of the Gujarat State Petroleum Corporation.

Anglo–Persian Oil Company
The original name of British Petroleum (BP), formed in 1909 and later known as the Anglo–Iranian Oil Company. This changed its name to British Petroleum in 1954.

Angola, offshore oil and gas production in
Angola is an emerging nation in terms of hydrocarbon production in its waters, including **deep-sea oil and gas** from wells drilled quite recently. Production of crude oil now stands at approximately 1 million **bbl** per day. At present most of the natural gas accompanying the oil is

either burnt off at a **flare** or re-injected into the well to bring about **enhanced oil recovery** or, at the Sanha field, stored under the sea for a period until a plant for **LNG** manufacture is in place. The Houston-based Marathon Oil is particularly active in exploration in Angolan waters. Its most recent success (mid 2005) is the **Gengibre well.**

By 2009 Angola will have a major refining facility at **Lobito**.

Antaramut-Kurtan-Dzoragukh (AKD)

Coal deposit in Armenia likely to enter use in the near future. Recent exploration has revealed that the deposit is much more extensive than previously realised and 31 Mt of **bituminous coal** are believed to be recoverable. The calorific value is high, about 33 MJ kg^{-1}. There is the disadvantage of a high sulphur content, around 3 %, and this will reduce the attractiveness of the coal for general-purpose use. However, Armenia is lacking in fuel resources and has recently suffered from an unreliable supply of gas from Russia. The AKD deposit therefore represents a significant resource.

Antarctic treaty

Though there is oil production in the Arctic, oil exploration in the Antarctic is expressly precluded by Antarctic Treaty of 1998. This is not likely to be reconsidered before 2048.

Anthracite

The most advanced coal along the **coalification** sequence. Anthracites have carbon contents up to about 95%. The country with the largest anthracite reserves is China. In North America, major deposits were discovered in 1762 in Pennsylvania and these are still amongst the world's largest anthracite reserves. In the UK, major anthracite deposits are to be found in Wales. As coalification takes its course, the differences between the **macerals** is reduced therefore petrographic analysis is less helpful for an anthracite than for, say, a **lignite**.

A quality low-ash anthracite such as **black diamond** has a calorific value of 32 MJ kg^{-1}. Being very low in volatiles, anthracites are not suitable for coking, as particles do not coalesce on heating. However, anthracites can be used as a metallurgical reductant in some applications as mined without subsequent carbonisation. Anthracite produces little smoke on combustion, which makes it suitable for domestic use. When coal fires in homes were more numerous than they now are, over 50% of the vast anthracite production of Pennsylvania was accounted for by domestic use. Use of anthracites in applications such as steam raising is usually seen as being wasteful except for the dust which has formed by breakage in handling. Even this however has the disadvantage that the hardness of anthracites necessitates more energy for grinding in preparation for utilisation than for coals of lower rank. Also the paucity of volatiles makes for difficulties with ignition of small particles. Nevertheless there has been significant usage of anthracite in this way, notably in Wales. Anthracite has a much lower propensity to spontaneous heating when stockpiled than lower rank coals do and fine particles of it are not a dust explosion hazard to the extent that fine particles of lower rank coals are.

Anthracites are not only a fuel but a raw material for the manufacture of carbon products including electrodes and absorbent powders and granules used for example to decolourise a solution of a chemical product prior to evaporation. Much of the anthracite produced worldwide at present is used in such applications.

Appleton, WI
Scene of the first **hydroelectricity** generation in 1882. Power for street lighting and for two local paper mills was provided.

API gravity
Index of the density of a liquid hydrocarbon, first used in the early days of the industry and still the most common means of expressing such densities:

$$\text{degrees API} = \frac{141.5}{\text{specific gravity at } 60°\text{F (water } = 1)} - 131.5$$

For example, a crude oil with a specific gravity at 60°F (= 15.6°C) of 0.85 has a density of 35 degrees API. Though the API (American Petroleum Institute) scale does find application to distillates and residues, its original formulation was for the classification of crudes. The more abundant a crude is in the lighter fractions, especially gasoline, the lower its specific gravity therefore the higher its density on the API scale, which incorporates the reciprocal of the actual density. Where light fractions are of the most value there is a positive correlation between quality and degrees API.

Aqueous film-forming foams
Made from water, a foaming compound and a surfactant, applications include spillages of leaked liquid fuels. The same performance criteria apply to these as to **protein foams**.

Aracoma mine, WV
Bituminous coal mine in the Appalachian region where two miners were killed in a fire in January 2006, the same month as the accident at **Sago Mine WV**.

Arco, Idaho
The first town to have street lighting from power generated using a nuclear fuel, in 1955. The generating facility, developed by the National Reactor Testing Station in Idaho, had an output of 5 MW. At Obninsk in Russia a nuclear electricity plant, also with 5 MW output, had come into operation about six months earlier.

Arctic ice shelf, recent contraction of
The Arctic ice shelf currently occupies an area of about 2 million square miles. According to satellite imagery and earlier shipping records, this is the smallest value for a century. During the relatively recent interval 1978 to 2000 the average area was closer to 3 million square miles. The immediate threat is to the people of the Inuit race, approximately 155 000 in number, who are fearful that one generation down the track their culture and way of life will have ceased.

The shrinkage of the Arctic shelf is very high on the agenda regarding greenhouse gas emissions. However, there is a fairly strong natural geophysical factor known as the Arctic oscillation, due to variations in the atmospheric pressure at sea level. To determine the relative importance of each may not be straightforward. Also, a cumulative effect of greenhouse gas emissions and the melting they cause is that solid ice and liquid water have significantly different emissivities and therefore differ in the extent to which they absorb solar radiation.

Arctic National Wildlife Refuge
Area currently protected from oil and gas exploration. At the time of writing, reconsideration is being discussed in places including US Congress.

Argyll field
The first oil and gas field in The North Sea to enter production, later renamed the Ardmore field. in 1992, after about 20 years of production, it was deemed uneconomic and operations ceased. It is however known that it still contains 21 million **bbl** of oil which ought to be accessible with advanced drilling technologies; development work is currently underway with a view to resuming production.

Arizona Clean Fuels Yuma

Once Arizona Clean Fuels Yuma complete their refinery in Yuma County, about 100 miles from Phoenix, it will be the first **grass roots refinery** built in the US for almost 30 years. It will be linked directly to a fuel terminal in Mexico and is expected to process chiefly Mexican oil into diesel, kerosene and gasoline at a rate of about 150 000 **bbl** per day. Arizona is currently dependent on California and Texas for such fuels and the new refinery is intended to eliminate such dependence.

Armada Platform

Platform for production of non-associated gas and condensate, situated in block 22/5 of the North Sea. Gas and condensate are separated at the platform and transferred separately to the Central Area Transmission System (CATS) facility in block 22/9 from which the condensate is sent to Cruden Bay, north of Aberdeen, and the gas to a terminal on the north eastern coast of England. The platform is operated by BP.

Aromatics

Unsaturated cyclic hydrocarbons containing one or more rings (usually benzene rings). Originally named for their distinctive odours. For example:

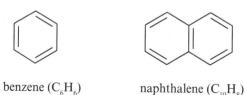

benzene (C_6H_6)   naphthalene ($C_{10}H_8$)

Arrhenius parameters

Quantities expressing the temperature dependence of a chemical reaction rate:

$$\text{rate} \propto A\exp\left(\frac{-E}{RT}\right)$$

where E (J mol^{-1}) is the activation energy, A (the units of which depend on those of the proportionality constant) is a constant dependent upon the particular reaction, R is the gas constant and T is the absolute temperature.

The expression A has its origin in collision theory of gases; indeed it is sometimes called the frequency factor and is related to (or in some cases actually equal to) the collision frequency of the gaseous molecules reacting. Reactions other than gaseous ones often have a temperature dependence of rate such that an expression having the same form as the Arrhenius expression applies. This is true of reactions of interest to fuel technologists, for example coal combustion, and the proportionality factor is often density (kg m^{-3}). The Arrhenius equation is then:

$$\text{rate} \propto \rho A \exp\left(\frac{-E}{RT}\right)$$

where ρ is density and A has units s^{-1}, the rate being in kg m^{-3} s^{-1}. The point made earlier that many reactions other than gaseous ones conform to the Arrhenius equation as expressed above could perhaps be worded with a different nuance: all chemical reactions are expected to conform to such a functional form for temperature dependence of rate. When, from experimental data, one is found which does not conform, it is labelled 'non-Arrhenius behaviour'. The implication of that term is that Arrhenius behaviour is what is generally observed. This is true not only in fuel technology and combustion chemistry but also, for example, in biochemistry.

When an Arrhenius expression is used for such things as coals, the parameters A and E may or may not have a mechanistic interpretation. It is fair to say that more commonly they do not, and are seen only as experimentally deduced parameters although very useful in the prediction of the behaviour of the substance to which they apply across a temperature range. Sometimes attempts are made to interpret experimentally determined A and E values on a mechanistic basis.

Two examples of such interpretation, among the many which can be read of in the research literature, are briefly mentioned. First, Arrhenius parameters for the combustion of a porous carbonaceous material might give some indication of the extent to which the reaction rate is controlled by diffusion of oxygen into the pores. Secondly, in wood combustion the **pyrolysis** which precedes and accompanies combustion and provides vapour-phase fuel might be resolvable into pyrolysis reactions on the part of the three constituents—cellulose, hemicellulose and lignin—of the wood with separate Arrhenius parameters for the pyrolysis of each.

Arthit field
Non-associated gas field in the Gulf of Thailand. A recently completed pipeline will convey gas from the field to southern Thailand. The estimated reserves are 55 billion m^3 of gas and production by 2010 is projected to be 14 million m^3 per day giving a life expectancy of about 10 years.

Arthur field
Shallow water gas field in the North Sea, only 30 miles out to sea from the Norfolk coast and one of the most recent North Sea fields to have come into production. There are now two production wells with a combined yield of up to 3.5 million m^3 of gas per day.

Ash
Solid residue following combustion, usually containing the inorganic components of a fuel after the organic components have been burned. See **bottom ash, fly ash**.

Asia, growth of nuclear generation of electricity in
Using uranium imported from Australia, Japan currently has a significant nuclear electricity generation. Recent events (including **Hurricane Katrina**) have not only caused oil prices to rise but also heightened world consciousness to the possible unreliability of such supplies (the *raison d'etre* of the **Strategic Petroleum Reserve**). Japan is promoting nuclear energy for itself and other Asian nations, for example by providing know-how for China to extend its nuclear electricity generation capability. There will of course be CO_2 reductions as a bonus. Indonesia and Vietnam are also poised for expansion of nuclear power.

Asmol
Trade name of a newly developed material for coating the outside of buried oil and gas pipelines to make them resistant to corrosion. It is organic, the starting material for its manufacture being heavy petroleum residue. Research and development indicate that once applied a coating of this substance will last 30 years, which might well exceed the period of service of the pipe.

Asphaltene
Large, complex, very high molecular weight bitumen compounds containing carbon, hydrogen, oxygen, nitrogen and sulphur, dissolved or

dispersed in crude oil and responsible for its colour. High asphaltene oils are extremely difficult to burn due to the stability of the bitumen.

Associated gas

Natural gas occurring with crude oil and requiring separation from it. It is distinct from 'non-associated gas' such as that from the **Morecambe Bay gas field**. Associated gas accounts for a significant proportion of the world's natural gas production. As an oil field furnishing associated gas ages the oil-to-gas proportion will change. All of the natural gas originating from Kuwait is associated gas. One very occasionally encounters documents in which the word 'associated' is applied to **liquefied petroleum gas (LPG)** removed from crude oil at the well instead of at the refinery. This terminology is best avoided, not only to avoid confusion but because it is not correct. The gas which becomes LPG was not 'associated' with the crude oil: it was part of the crude oil itself.

ASTM International

Founded in 1898 as the American Society for Testing of Materials, ASTM International is a voluntary organisation that produces technical standards for materials, products, systems and services.

Athabasca sands

A source of **bitumen** fuel, first discovered in 1875 and a source of hydrocarbons for western Canada. The tar sands of the Athabasca region of Alberta potentially contain more oil than the proven crude reserves of Saudi Arabia. The term 'Athabasca sands' has acquired a broadened meaning and takes in tar sands from four regions of Alberta: Athabasca, Wabasca, Cold Lake and Peace River.

The Alberta provincial government reports that there are 1.6 trillion **bbl** of bitumen recoverable. The bitumen occurs with sand, from which it has to be separated by treatment with water. Bitumen separated from the water is then diluted for pipeline transfer to a treatment plant where it is hydrogenated to **syncrude** which yields on distillation products corresponding to the fractions from crude oil including gasoline. Gasoline so obtained has been available in Canada since the 1950s and current activity is as strong as ever, **Shell** being a major participant largely at its **Scotford Upgrader** facility. At the present time, production of liquid fuels from tar sands in Alberta is of the order of half a million bbl per day, approaching a quarter of Canada's liquid fuel requirements.

Although tar sands do exist in a number of states of the US including California there is little utilisation of them at present.

Atomisation

Process of dispersion of a liquid fuel into fine droplets before admittance to a **burner**. Such droplets have diameters tens or hundreds of a micron. Across a wide range of burner designs and sizes, a very common type of atomiser is the spray nozzle. There are two basic forms of this. In one, the liquid alone is passed through a nozzle and the kinetic energy gain is sufficient to break up the liquid into droplets. In the other, the liquid is mixed either with air or with steam before nozzle entry and the kinetic energy effect is enhanced by swirl. Another widely used type of atomiser is the rotating cup, in which dispersion is achieved by passing the liquid fuel through a cup rotating at a rate of several thousand rpm.

Aurora, NC

The site of a proposed plant for **ethanol** production which, when in comes into operation, will be the first such facility on the US east coast. Its target production rate is 114 million gallons per year.

Austria, energy scene in

This country has significant amounts of crude oil and of natural gas, proven reserves of 85 million **bbl** and 25 **bcm** respectively. There is a major oil refinery at Schwechat near Vienna. **Lignite** is produced at about 1200 tonnes per year and used for power generation. Higher rank coal is imported from Russia, Poland and the Czech Republic. Well over half of Austria's electricity is produced hydroelectrically. The country has much potential for the production of wood fuel and developments on that front are expected.

Autogas

Synonym for **liquefied petroleum gas** where it is used as a fuel for road vehicles.

Avgas

Petroleum distillate in the gasoline boiling range, used for small aircraft which have piston engines.

Avonmouth

English town in the south west long associated with the hydrocarbon industry, there having been a major explosion there as long ago as 1951 when two persons were killed. Currently, it has a terminal for **liquefied petroleum gas (LPG)** brought by rail from **Wytch farm**. The LPG storage containers at Avonmouth are spherical and range in the amount of LPG they can hold, the biggest holding up to 2000 tonnes.

Azeri field
Newly exploited oil field in the Caspian Sea operated by BP. The oil is carried by the **Baku-Tbilisi-Ceyhan Pipeline** to the Mediterranean coast. 2005 production: 93 000 **bbl** per day.

B

Bagasse
The cellulosic residue from sugar cane often used as a fuel in the manufacture of sugar. Its calorific value when air-dried is around 15 MJ kg^{-1}. Its storage sometimes results in spontaneous heating exacerbated by the fact that the material is, following sucrose extraction, well above ambient temperature when stockpiled. There is some current interest in carbonising bagasse to make chars.

Baghdad battery
At 2000 years old this is thought to be the first known **battery** and was possibly used for electroplating. Discovered just outside Baghdad, it consists of a clay jar with asphalt stopper through which sits an iron rod surrounded by a copper cylinder. When the jar is filled with an electrolyte, it has an electrical potential of about 1.1 V.

Bahrain, oil production by
The first nation on what is known as the 'Arab side' of the Persian Gulf to discover oil, in 1932, Bahrain is now a very small player in world terms producing 38 000 **bbl** per day which is taken to the country's one refinery which is currently being modernised. **NGL** from the **associated natural gas** is however exported.

Baja California, Mexico, liquefied natural gas (LNG) reception at
Construction is underway of an **LNG** terminal at Baja California which

will receive LNG from sources including Australia. There will be a pipeline structure to take the LNG once converted back to gas at an offshore vaporising unit, to northern Baja California as well as to California USA. The expected rate of supply from the terminal, which will enter service in 2007, is 19 million m³ of gas per day.

Bakassi Peninsula
Oil-rich region of western Africa. Rights to the oil are disputed, Nigeria and Cameroon each asserting a claim. Armed confrontations have sometimes resulted.

Baker Report
Following the 2005 explosion at the BP refinery in Texas City in which there were 15 deaths and 180 injuries, former US Secretary of State James Baker prepared a report covering the accident and the safety of BP's operations more widely in the US. The report extends to 374 pages, so an attempt to summarise it would inevitably misrepresent it. One of the points emphasised in the Report and subsequently commented upon in *The Times* will however be outlined. It appears that BP in the US had a very impressive 'personal safety' record in recent years. Personal safety measures in BP were described as 'obsessive', and an example given was of the imposition of speed limits below the legal ones on BP vehicles. The Baker Report makes the point that before the Texas City accident a good record in 'personal safety' had engendered an unjustifiable confidence in 'process safety'.

Bakersfield, CA
Scene of electricity generation at 0.5 MW level by **photovoltaic cells**. The power generated is used at a nearby oil field. The activity is part of Chevron's renewables programme.

Bakerton, PA
The scene of dismantlement of a pile of **boney-bituminous coal** waste, which has been there for nearly a century. The Houston based company Reliant Energy take fuel from the pile by bulldozer and burn it in a **fluidised bed** at a nearby power station, incorporating **limestone** to reduce sulphur emissions. Electricity is produced at about 500 MW, and the necessary coal consumption is such that the pile should be cleared after about four years at which stage the area can be landscaped. Reliant Energy obtain the fuel at no cost, an important factor in making the undertaking viable. A relevant point which does not

favour the operation is that Pennsylvania does not need extra electricity: it already exports more electricity than any other state of the US.

Baku-Tbilisi-Ceyhan Pipeline

Newly completed after about ten years of work, this pipeline conveys oil from the **Caspian Sea** to the Mediterranean. It is 1600 km in length and will hold 10 million **bbl** of crude oil. It was built by a consortium of companies, BP having a 30% stake. It is widely reported that the incentive for its construction was elimination of the dependence of the West on reserves of oil in the Persian Gulf region. There have of course been opponents to the construction of the pipeline on environmental and political grounds, and one particularly vitriolic commentary on the Internet has coined the term 'Pipelineistan'.

Oil transferred along the Baku-Tbilisi-Ceyhan Pipeline which is destined for markets in the Far East (or possibly the west side of the US) is passed along the **Eilat-Ashkelon Pipeline** for the next stage of its journey.

Balgzand-Bacton line

The first natural gas pipeline between the Netherlands and the UK, commencing operations in the Netherlands-to-UK direction in late 2006. Its capacity is 15 **bcm** per year of gas. In 2005 the UK became a net importer of natural gas for the first time, initially receiving most of its imported gas from Norway.

Ballast, in oil tankers

Once an oil tanker has offloaded its payload of crude oil it takes on seawater as ballast. Some tankers also contain tanks for ballast only to be filled, partially or wholly, to stabilise the tanker as sea and weather conditions change. Ballast from previously emptied crude oil tanks is 'unsegregated ballast' while that from the tanks for ballast only is 'segregated ballast'. Unsegregated ballast contains about 1% oil and has to be treated before re-admittance to the sea. With segregated ballast there is the possibility that if it was taken on at a location distant from the terminal, sea life not indigenous to the waters close to the terminal will accompany it, meaning release of the ballast into the sea at the terminal could adversely affect the ecosystem there.

Baltic pipeline system

Completed in 2005, this will convey annually quantities of the order of 10^8 **bbl**. At the terminal at the Russian port of Primorsk only vessels of the **double-hulled tanker** type will be used and very high standards of safety and environmental protection are expected.

Baltic Power Station

Fuelled entirely by shale products from Estonian sources, this facility has for 50 years supplied power at about 1.5 GW. Its importance is declining; although Estonia is a leader in shale oil production, power generation for Estonia itself from shale-derived fuels has become less attractive in the early years of the 21st century. It is expected that some of the shale oil exported from Estonia will continue to be used in power generation.

Baltimore, MD, early street illumination in

Baltimore was the first US city to have street lighting, in 1816. The fuel was **retort coal gas**.

Bangladesh, natural gas utilisation in

Though having almost negligible oil and very little coal, Bangladesh has large reserves of natural gas. It is also good quality natural gas, methane content in the range 93–99%. The country obtains about 70 % of its commercial energy needs from natural gas. The first discovery of natural gas in Bangladesh was in 1955 (when Bangladesh was East Pakistan) and usage began in 1961. There are twelve fields currently in operation, with an annual production currently of the order 10 **bcm**. Since the country has no oil to speak of, the **natural gas liquids** obtainable from the gas are a significant bonus. Exploration for offshore oil is also underway as there have been promising results from surveys.

Barbados, oil and gas in

This Caribbean country has modest oil reserves, the 2003 production being 1000 **bbl** per day. It is sent to Trinidad for refining and then returned to Barbados. Barbados also produces about 30 million m^3 of natural gas per year, sufficient to meet local demand.

Barge

Type of vessel used for offshore hydrocarbon production in which there is no production plant above the sea. Hydrocarbon is taken from wells at the seabed by pipeline to a barge berthed onshore some distance from the wells. The barge may simply collect or, as in the case of the **Hammerfest LNG** project, incorporate a processing plant. A barge can be towed from one scene of such activity to another thereby enabling **stranded gas** to be accessed.

Barnett Shale deposit

Onshore deposit in Texas currently producing 600 **bcm** of natural gas per month. The structure is such that in a well at the Barnett deposit descending to 7000 feet there is a layer of gas of about 550 feet and penetration of a layer of shale, trapped by **limestone**, will be required to access the gas. This is the origin of the name of the deposit. Barnett is the most productive gas field in Texas at the time of writing.

Barrel

Unit of volume of crude oil for pricing purposes, equivalent to 159 litres or 42 US gallons.

Barrel of crude, prices of since WW2

The prices in this entry actually relate to crude from one particular source, the Illinois Basin. Production there continues over 100 years after it commenced, 3.2 billion **bbl** having been produced to date with 4 billion bbl remaining. The choice of this crude as a representative for the purpose of this discussion is quite reasonable.

The price in mid 1946 was $1.72 per bbl, and this rose to a little over $2 over the next twelve months and to $3.02 by mid 1953. By the end of 1957 $3.00 would buy a barrel of this crude and in August 1960, the month before the formation of **OPEC**, the price was exactly $3.00 rising only to $3.35 by the end of the 1960s. OPEC was by that time becoming stronger and more influential in setting prices, having acquired several more member states. As noted in two other entries of this volume, OPEC response to US military presence in Israel in 1973 caused a sharp rise in the price of oil, and Illinois basin crude rose in price by 45% from $3.60 to $5.20 between April and December that year. A price of $5.20 was sustained during 1975 and this had risen to a few cents below $6 by the end of the decade. Over the 1980s the price rose from $6.34 to $19.04 and, after dropping to around $12 for a period in the late 1990s, approached $25 in the year 2000. The average price over the first eight months of 2005 was $49.71 and this rose to about $68 during September.

Barrels per stream day (BPSD)

The capacity or performance of a refinery will usually be expressed in **bbl** per day on the basis that the refinery operates 24 hours a day for almost the whole year, three weeks per year of 'down time' being about the norm. At refineries where operation is for any reason intermittent the appropriate measure is bbl per *stream* day, abbreviation BPSD.

Base gas

Also known as cushion gas, this is the minimum amount of natural gas that must be stored to keep production and deliverability at the required rate and pressure to ensure continuous operation.

Basis

Most analyses on coals are carried out on air-dried (ad) samples (so includes inherent moisture but not surface moisture), but other bases are often used.

These are: as received (ar), which takes into account fixed carbon, volatile matter, ash and moisture; dry (db), which takes into account the fixed carbon, volatile matter and ash but no moisture; dry ash free (daf), which only accounts for fixed carbon and volatile matter; and dry mineral matter free (dmmf), which takes account of the fixed carbon and volatile organic matter. A dmmf basis excludes moisture, ash and volatile mineral matter content.

There is one further basis: rom (run-of-mine), which is used for samples taken directly at the mine and so does not include further effects from transport and handling.

Baseline emissions

The average emissions over a defined period of time, subject to no changes that would affect emissions levels (such as the installation of a flue gas **desulpurisation** plant in the monitored area, causing emissions to be reduced).

Baseload

The load a power station must continuously provide to satisfy a minimum requirement (cf **peak load**). This is often provided by a baseload plant, operating continuously at a constant rate, to meet all or part of the minimum system load requirements.

Basestock

Material in preponderance – 75 to 85% – in a lubricating oil. It might consist of residual material from refineries having had no further treatment, or such material having had some degree of treatment with hydrogen. In going from basestock to saleable lubricating oil, further hydrocarbon (by way of blendstock) is used to adjust the properties including the very important one, for lubricants, of viscosity. The viscosity of a hydrocarbon lubricant varies very sharply with temperature, another factor to be considered in the manufacture of lubricants. Basestock for **synthetic motor oils** is **Alpha olefins**.

Batman refinery

The oldest of the five refineries in Turkey, operated by Turkish Petroleum Refineries Corporation. It began operations in 1955 and, after a number of enlargements, can now process 7 million **bbl** per year.

Recently a significant part of Turkey's oil refining facilities have been taken over by **Shell**. An Indian oil company bidding for the same facilities was unsuccessful.

Battery

A direct current voltage source to convert chemical energy into electrical energy through two or more electrochemical cells connected in series or parallel. Each cell consists of a vessel containing a positive and a negative electrode, and electrolyte held between the two. The 'AA batteries' used in many devices such as digital cameras are, in fact, single cells. Invented by Alessandro Volta in 1800 (see **Baghdad battery**) and subsequently exploited by many scientists such as Faraday.

Baumé scale

Original name of the **API gravity** scale and the current name of another closely related scale which uses a numerical value of 140 in place of the 141.5 in the API gravity scale. Differences in the Baumé and API values will be about a tenth of a degree for a heavy fuel oil rising to approaching a degree for a gasoline.

Bay Front, Wisconsin

The scene of thermal generation of electricity at 75 MW using a remarkable variety of fuels. Sometimes conventional fuels—coal or wood—are used but there is the means to utilise much less orthodox fuels as and when they become available. These include rubber tyre waste, having previously been shredded, rubber waste from other sources and also waste floor covering materials. The strength of this enterprise is its flexibility in terms of fuel use.

Bayport

Projected oil terminal for the Gulf of Mexico, which will have a capacity of 1 million **twenty-foot equivalent units (TEU)**.

Baytown, TX

Scene of the largest refinery in the US, operated by Exxon Mobil; its capacity is 557 000 **bbl** per day. Temporary complete shutdown occurred there as a result of **Hurricane Katrina**.

25

Bayway refinery
Situated in New York Harbour and operated by ConocoPhillips, this facility is capable of processing 238 thousand **bbl** of crude oil per day. Imported crudes from the North Sea, Canada and Nigeria are received there and the refined products distributed along the east coast of the US. There is also a polypropylene plant at the site.

Bayway refinery operates in co-ordination with the Trainer refinery in Philadelphia, also operated by Conoco-Phillips. Some crude received in New York Harbour at the Bayway refinery is diverted to Trainer. Refined products from Trainer are distributed only within Pennsylvania, New Jersey and New York.

bbl
Abbreviation for the unit **Barrel.**

bcm
Abbreviation for one billion cubic metres i.e. 1×10^9 m^3.

BCURA probe
A widely used device, developed by British Coal Utilisation Research Association (no longer in existence), for measuring the concentration of **fly ash** in gas from the **combustion** of **pulverised fuel**. The gas is drawn into a cyclone and from there to a receiver. Special conditions have to be applied to the receiver so that the influence of the edge effects on the gas flow are accounted for. Therefore, the measured concentration of fly ash remains constant no matter the distance from the receiver.

Bear Creek, PA
Scene of a major wind farm in Pennsylvania, expected to have 12 turbines each of 2 MW output, when complete. Also in Pennsylvania is the Jersey-Atlantic Wind farm with five turbines each of 1.5 MW output.

Beatrice Wind Farm
Offshore wind farm in the Firth of Moray in northeast Scotland. A single demonstrator turbine was installed at the commencement of operation in summer 2006. There will be expansion on the basis of performance monitoring of the initial turbine. This and onshore wind generation in Scotland will help to meet the country's target of 40% of its electricity from renewable sources by 2020.

Beijing, atmospheric reform in

Partly because it is hosting the Games of the 29th Olympiad in 2008, this city is taking quite extensive measures to improve its air quality. Between the time of writing and 2007, the coal utilisation in Beijing will have been reduced by a third. Natural gas consumption will consequently increase and a new pipeline to the city is being installed to provide for this. In certain areas, factories will be required either to lower their emissions or move elsewhere. Restrictions on access to parts of the city by drivers of privately owned vehicles apply and vehicles older than fifteen years will be compulsorily taken off the roads. There will also be planting of trees. Sinopec are using the **S Zorb™ desulphurisation** procedure at refineries in cities including Beijing.

Belgium, demise of the coal industry in

Belgium perhaps provides the most striking example of the decline in coal utilisation over the last few decades. The country is in fact quite well endowed with **bituminous coal**. Around the mid 20th century (when the explosion at **Marcinelle** occurred), Belguim was producing 20–25 Mt yr^{-1}, significantly higher than the Canadian production and about the same as the South African production at that time. There has however been no coal mining in Belgium since 1992: coal has been replaced by nuclear fuels and imported conventional fuels.

Belize, oil production in

For very many years exploration for oil in Belize met with disappointment, fifty exploration wells having revealed either oil which was not economically accessible or no oil at all. Recently 500 **bbl** per day have been produced at a newly developed well: the crude is sweet and light. Clearly the present production is negligible but there is promise of greater yields and in Belize, population just over a quarter of a million, even 50 000 bbl per day would make a major difference to the oil supply and demand situation. This is the target to which the developers of the field are currently aiming. Incentive is provided by the current high prices of oil which could affect national economies far stronger than that of Belize!

Beloyarsk, Russia

Scene of construction of what when complete will be the world's largest **fast reactor** for electricity production. Having regard to the fact that that one of the constituents **Mixed oxide (MOX)** fuel is uranium depleted in the 235 isotope therefore enriched in the (already dominant)

238 isotope, MOX is a suitable fuel for a fast reactor and will in fact be used at the Beloyarsk facility. There is already such use of MOX France.

Benchmark crudes

When a particular crude oil is being traded, its quality, in terms of such properties as **API gravity** and sulphur content, determines its price. This requires a benchmark against which the price of the oil being traded can be assessed. There are a number of such benchmark crudes and these are briefly described in the table below.

At any one time there can be a difference of ≈ 15% between the highest and lowest reference price per **barrel** on the basis of the standards in the table above. One has to bear in mind that oil prices respond to events which might not on first consideration seem relevant. For example, **OPEC** price was affected by the death of King Fahd of Saudi Arabia.

Benchmark crude	*Description*
Brent	The world benchmark, Brent is a blend of North Sea oils having an API gravity of 38.3 degrees and a sulphur content of about 0.37%. It is therefore both 'light' and 'sweet'.
West Texas Intermediate	The US benchmark, having an API gravity of 39.6 degrees and a sulphur content of about 0.24%. It is therefore lighter and sweeter than Brent.
OPEC Basket Price	OPEC countries do not use a specified blend as a pricing standard but a hypothetical 'basket' of seven as-received crudes, from: Algeria, Indonesia, Nigeria, Saudi Arabia, Dubai, Venezuela and Mexico (which is not in OPEC). Brent and WestTexas Intermediate are both engineered to be sweet and light, but the 'basket' used by OPEC is, when averaged across the seven constituents, less sweet and less light than either.
Dubai crude	The benchmark used for the purposes of pricing Middle East crudes for the Far East market, especially those exported to South Korea.
Tapis and Minas	The reference for crudes originating from the Far East, sometimes referred to as 'Asian sweet crude'.

Beneficiation

The adding of value to a solid fuel in its raw form by processing, for example by drying, by removal of some of the ash-forming constituents or by briquetting.

Benzol(e)

Light oil from a by-product coke oven incorporating the **Benzene, Toluene, Xylenes (BTX)** yield.

Bergius process

A means of making liquid fuel from **coal** by hydrogenation. The coal in pulverised form is made into a paste with oil and reacted with hydrogen at about 450°C and pressures in the neighbourhood of 400 bar. A great deal of the aviation fuel used by the Luftwaffe during WW2 was made this way using low-rank coal (**braunkohle**). There is at the present time a great deal of activity in the manufacture of liquid fuels from such coals but with hydrogen-donating solvents, the best known example of which is **tetralin**, rather than with hydrogen gas. Related research and development is largely concerned with catalysis as indeed was the case early in the technology. Bergius and Karl Bosch shared the 1931 Nobel Prize for Chemistry for work on 'high pressure processes', and Bosch's contribution was finding a suitable catalyst. He systematically tested many hundred possible catalysts.

Bergville

Town in the Kwazulu province of South Africa. In the nearby Drakensburg mountains there is a facility for **pumped storage of electricty**. The capacity is 1 GW.

Bermuda, energy scene in

This island community of population 65 000, still under the British Crown, obtains its electricity from gas turbines and **genset**s operated by the Bermuda Electric Light Company (BELCO). Demand seldom exceeds 100 MW, and the gensets provide the base load, the gas turbines providing supplementary power as necessary. Both devices use imported petroleum-derived fuels. There is no natural gas supply to Bermuda but significant amounts of **liquefied petroleum gas** are imported as a general-purpose fuel. In 2005 there was a fire at a BELCO transmission facility resulting in total power loss. Power was restored to 90% of the residents the same day.

Bideford Dolphin

Semi-submersible having a particularly interesting history. It was built in Japan and saw service primarily in the North Sea, but also in the Mediterranean and off the Portuguese coast, until 1986. At that point it was put into storage at a Norwegian Fjord for 11 years until being towed to Stavanger, Norway, for the first stage of a rebuild. The rebuild was completed in Belfast and the Bideford Dolphin re-entered service in the North Sea in 1999. It has recently been involved in the drilling of production wells in the Norwegian Sector of the North Sea and will shortly be used in the replacement of the **Christmas tree**—the interface of the well and platform pipe work—at an existing platform. Bideford Dolphin in its rebuilt form can accommodate up to 88 persons.

Big cat
Term applied in the oil industry to a field under exploration which is expected to contain amounts of oil in excess of 100 million **bbl**. There are two such in Nigeria at the present time, being explored by **Shell**.

Big Foot field, TX
First discovered by Shell in 1949, this field has yielded 22 million **bbl** of crude oil per day for 50 years. The field is situated southwest of San Antonio and is believed to have reserves in excess of 5 million bbl remaining.

Big Sandy River, WV
The scene of over 12 coal terminals one of which, the Big Sandy Terminal, has an annual throughput of 7 million US tons and storage facilities for 250 000 tons. It operates on a 24-hour basis for five days in the week and has facilities for blending coals. In the addition of delivery charges to a **Central Appalachian (CAPP)** price for a coal the distance from the Big Sandy River to the ultimate destination of the coal is used as a basis for calculation.

Bina Hill, Guyana
Scene of electricity generation by **photovoltaic cells** for the benefit of a native American settlement which previously relied on **gensets** for power. The photovoltaic equipment was manufactured in Germany, and the power obtained is used to provide lighting and PCs for schools.

Bintulu
Coastal town in east Malaysia and the site of the world's largest **liquefied natural gas (LNG)** shipping terminal.

Bio Bio River, Chile

The scene of major activity in **hydroelectricity**. The hydroelectric facility, which is about 300 miles south of the capital Santiago, provides 570 MW of electricity, raising the previous power generating capacity of Chile by 18%. There are several other hydroelectric installations in Chile.

Biodiesel

Plant oils used as substitutes for diesel in compression-ignition engines and obtained from the crushing of seeds. Significant development of this took place in India in the 1980s where Honne oil was investigated as a diesel substitute with promising results. There is continuing activity in such fuels encouraged, in certain parts of the US, by tax incentives for users of them. Adjusting biodiesel fuel to required properties is possible, for example by viscosity blending with animal fat. Biodiesel fuels are non-toxic and biodegradable, two enormous advantages over conventional (mineral) diesel. Biodiesel fuels also hold promise for **genset** operation. There is significant investigation of their potential as a fuel for ships in the **BioMer Project** currently under way in Canada.

Equivalence of biodiesel and mineral diesel requires that the oxygenated organics in the biodiesel, in particular the esters, have a **combustion** reactivity comparable to that of unsubstituted hydrocarbons up to about the C_{22} in the mineral diesel. To be brought to within specification, a biodiesel might need to be modified in composition by raising its ester content. This is routine practice when biodiesel is made from **hemp** seed. Being plant derived, biodiesels tend to contain some water, which is a disadvantage in use. There are agents including **Acusorb** for dewatering biodiesel.

Biogene

One definition of **municipal solid waste (MSW)** from the viewpoint of its use as a fuel is that it is 50% biogene and 50% non-biogene. Some compilations of **renewable energy** resources divide the total weight of MSW by two and so report reserves according to this convention.

Biomass

Organic matter of non-geological origin that can be used for energy production. Biomass can include commercial wastes, **municipal solid waste**, wood and plant residues, and **energy crops**.

BioMer Project

Scheme whereby **biodiesel** will, for a specified trial period, be used to power a fleet of 12 vessels in use in parts of Ontario including the Port of Montreal and the Lachine Canal. Eleven of the vessels will be powered on neat biodiesel and the other on a blend of biodiesel and conventional fuel, the requirement over the duration of the trial being about a quarter of a million litres of biodiesel.

Biorefinery

Proposed plant for obtaining hydrocarbon fuels from sugars, starch and cellulose. Alkanes in the range C_7 to C_{15} have, in exploratory work on the laboratory scale at the University of Wisconsin, Madison, been yielded from starting materials including newly harvested crops and agricultural waste. The sequence of reactions takes place in the aqueous phase, from an initial breakdown of the carbohydrates to carbonyl compounds. A catalyst is used in the synthesis of the carbon structures required in the final product and there is, on the basis of the lab scale work, promise of entirely sulphur-free gasoline and diesel fuels by this means.

Bird life, threat to from oil spills

Most readers will have seen tragic images of birds covered with oil spilled along a coastline. What is not widely realised is that very small amounts of oil, unnoticeable except by professional examination, can be fatal. A bird relies on air between its feathers for insulation and one or two drops of oil can, by surface tension effects, prevent the feathers from functioning properly, causing hypothermia.

Bit

Tool used to drill an oil well. It is attached to the drill pipe which rotates. The pipe/bit assembly is called the drilling string and will use a **drilling fluid**. A bit might be fabricated from a suitable alloy with teeth made of a cutting material such as tungsten carbide, or the entire bit might be made from industrial diamond. There are numerous designs and configurations of bits which have been developed in the industry over the years. As drilling of a well takes its course, bits progressively smaller in diameter are used. The first stage drilling, known as 'spudding in', possibly using a bit as large as two feet in diameter. When use of a bit of a particular diameter ceases, a metal casing will be inserted which has an outer diameter somewhat smaller than that of the bit so that there is an annulus between the drilled hole and the casing which can be filled with cement.

Bitumen

When bitumen occurs alone it can be seen as an extra heavy crude. For example, Venezuela has proven reserves of about 75 billion **bbl heavy crude** and 35.7 billion bbl of extra heavy crude a.k.a. bitumen. When it occurs in association with sand it is known as tar sands or, if it originates in Alberta Canada, as **Athabasca sands**. It is important to note that bitumen is *not* the same as kerogen, the organic component of shale.

Typically bitumen will have an API gravity of around 9 degrees and for liquid fuel production this needs to be adjusted to a value of around 30 degrees. The bitumen is blended with a light petroleum fraction, or possibly with natural gas condensate, so it can be pumped to an upgrading facility and hydro-treated to give **syncrude** which can then be refined. The Venezuelan bitumen referred to above is in the Orinoco belt in the east of the country. Syncrude from it is produced at a rate of 450 000 bbl per day and refined within Venezuela, much of the refined material being exported to countries including the US. Some of the bitumen produced in Venezuela is not refined at all but converted to **Orimulsion®**.

Bituminous coal

Coal well advanced along the coalification sequence having a carbon content in the approximate range 70–80% dry basis and a calorific value, depending on the moisture and ash contents, of about 30 MJ kg^{-1}. Uses include direct combustion, carbonisation to make **coke** and **gasification** to make **retort coal gas** or **blue water gas**. Bituminous coal has not only served industry for generations but played an important part in industrialisation itself by providing a material for the manufacture of coke subsequently used to make iron and steel for the construction of machinery. A region-by-region discussion of bituminous coal occurrence and utilisation is given in the table below. Only areas producing more than 100 Mt yr^{-1} feature in the table. Some of the production figures, notably those for China and for the former Soviet Union, are for **hard coal** not for bituminous only.

The table below is comprehensive in that all countries with a production in excess of 100 Mt yr^{-1} feature in it. Canada is a notable absentee. Its current annual production of coal of all ranks is about 73 Mt. That is not to say however that the coal industry in Canada is static. Exports recently began of bituminous coal from British Columbia to the Far East. These coals are suitable for **coke** manufacture and their ultimate destination will be the Japanese and Korean steel industries. The **UK coal industry** is presently such that it does not merit inclusion in the table by quite a long way. Bituminous coals continue as one of the world's staple fuels.

Country	Background to utilisation	Current production (Mt yr^{-1})	Comments
China	Long recognised as one of the world's most coal-rich countries.	1326	Mining practice in parts of China has become very lax and accidents are common (e.g. Liaoning province, February 2005, when over 200 were killed). Even so, China is currently the world's fifth largest importer of coal.
USA	Major deposits of bituminous coal in: AL, AR, CO, IL, IN, KS, KY, MD, MT, NM, OH, OK, PA, TN, UT, VA, WA, WV, WY. The first discovery of bituminous coal in what was later to become the US was in Virginia in 1701. About 90 years ago annual production $\approx$400 Mt. Major uses were power generation, **coke** production and **town gas** production.	$\approx$800	Production falls slightly short of demand and there are some exports of coal to the US, although 90 years ago the US was exporting *to* Canada and elsewhere.
India	Major bituminous coal fields at Bengal, Bihar and Madya-Pradesh. Mid-1960s production $\approx$30 Mt yr^{-1}.	340	The high ash content of some Indian bituminous coals reduces their market value.
Australia	Deposits of bituminous coal in New South Wales and Queensland. In 1920 annual production was 11 Mt.	270	Less than 3% of the US production in 1920 rising to about 14% of the US production in 2005.

Country	Background to utilisation	Current production (Mt yr⁻¹)	Comments
South Africa	Reserves of bituminous coal in Kwazulu-Natal, the Transvaal (including the part now known as the **Mpumalanga Province**), Orange Free State (scene of the **Coalbrook** disaster) and Cape Province.	223	South Africa is a major exporter of bituminous coal, most of it passing through the terminal at **Richards Bay**
Former Soviet Union	A paucity of statistical information for pre-1990s. Germany, Poland, the former Soviet Union and the Ukraine account for over 95% of the current European total production of bituminous coal. Poland produces about 100 Mt yr⁻¹. The Ukraine produces 83 Mt yr⁻¹.	160	The former Soviet Union is second only to the US in the magnitude of its proven coal reserves.
Indonesia	Vast amounts of coal in Indonesia but major utilisation only in relatively recent years.	≈ 100	Currently much foreign investment in coal mining in Indonesia by Australia, the US and Canada. Indonesia also consumes large amounts of **peat** in energy production.

Bjarnarflag, Iceland

The scene of one of several **geothermal electricity** plants in that country, with a combined capacity of 170 MW accounting for about 16% of the total power requirement. Iceland has strong geothermal reserves and other applications include direct industrial heating.

Blackdamp

Term applied to gas released during mining operations, capable of extinguishing the flame of a safety lamp. It is principally carbon dioxide.

Black diamond

High-quality anthracite mined in Wales, having an ash content of not more than 4%. It is supplied by Celtic Energy in particle size ranges specified by the purchaser. It is premium quality coal and sells in smaller quantities than other anthracites which Celtic Energy can supply.

Black liquor

By-product of paper manufacture suitable, after concentrating, for fuel use. Once the cellulose has been removed from wood in order to prepare pulp the residue which remains comprises the lignin from the wood suspended in water. This is too low in solids content to burn, but concentration to about 75% solids by evaporation of water gives a material having a **calorific value** of about 9 MJ kg^{-1} known as black liquor which can be burnt and the heat recovered.

Although black liquor is usually seen as an in-house fuel for the pulp and paper industry there is often some to spare, and compilations of utilisation of biomass fuels for a particular country might well include black liquor. There is increasing interest in the gasification of black liquor. It can also be fully dried and then carbonised.

Black start

The ability of a cold power plant to become fully operational through use of a dedicated auxiliary source that does not require an external power system.

Black Sunday

In the early 1980s there was major investment in developing the shale oil reserves of western Colorado. This was because of rises in the price of conventional crude oil, a factor possibly making shale-derived crude oil competitive. Then the price of conventional crude oil dipped dramatically and on Sunday May 2nd 1982, which became known as Black Sunday, one of the major investors in the western Colorado undertaking withdrew: the others followed. One of the many consequences was the loss of 10 000 jobs and bankruptcies of several local companies who had serviced the development.

Black Thunder mine

Mine in Wyoming yielding **sub-bituminous coal** low in sulphur, much of which is used in power generation. The mine is one of the most productive in North America. Black Thunder coal as mined contains 25–30% moisture and on burning yields an ash residue of about 5%

the coal weight. The coal (like many lower rank coals) is susceptible to spontaneous heating in storage, especially under humid conditions.

Blast furnace

The blast furnace is the most widely used technology for producing pig iron. The furnace is continually charged at the top with burden: iron ore, coke and **limestone** flux. Metallic iron is produced by the reduction of the iron ore by **carbon monoxide** gases formed inside the furnace at temperatures in excess of 2000°C and pressures of 500 kPa (5 atmospheres):

$$Fe_2O_{3(s)} + 3CO_{(g)} \rightarrow 2Fe_{(s)} + 3CO_{2(g)}$$

Hydrogen, which is more effective at reducing iron ore than carbon monoxide, is generated by the reaction of water vapour introduced with the blast two-thirds down the furnace, either as steam or from the coal injected into the blast with hot coke:

$$C_{(s)} + H_2O_{(g)} \rightarrow H_{2(g)} + CO_{(g)}$$

Impurities in the liquid iron are removed by the slag which floats on the surface of the metal and can be removed separately.

Due to increasing cost and decreasing availability of **metallurgical coke**, coal is increasingly being used to replace coke in the blast furnace. As well as being half the price of coke, coal injection allows the temperatures to be more easily controlled and increases the amount of hydrogen in the area of the blast.

Blast furnace gas

The effluent gas from a blast furnace containing enough **carbon monoxide** to make it flammable on a suitable **burner**. It is actually a form of **producer gas**. Dual-fuel combustion of blast furnace gas with a richer fuel such as fuel oil in droplet form is possible. It is toxic however, and over the years there have been many deaths within the iron and steel industry due to this. One of the most recent fatal accidents at a blast furnace was at **Port Talbot**.

BLEVE

See **Boiling liquid expanding vapour explosion**.

Block

Unit of offshore area for the purpose of the issue of licences to oil/gas companies. For this purpose, waters in the UK Continental Shelf are divided in quadrants of 1° latitude by 1° longitude and the first of the numerals assigned to a block will be the number of the quadrant. For example, blocks 20/10b, 21/6a, 20/15a and 21/11b, which together feature on a licence recently issued to Premier Oil and Reach Exploration jointly, are in adjacent quadrants 20 and 21 which are in fact out to sea from the Moray coast in the north of Scotland and are referred to as 'Moray Firth blocks'. As a further example, the 53rd quadrant is out to sea from the East Anglia coast and takes in the **Arthur Field** which is in fact in block 53/02. Moving to the west side of the land-mass, quadrant 110 is off the Lancashire coast and blocks in region of the **Morecambe Bay gas field** have numbers 110/***. The quadrant numbers from 1–10 encircle the Shetland Islands and some of the highest numbers, greater than 200, also occur in this region. For example the **Clair field**, which is only 75 km west of Shetland, is in quadrant number 206.

Blue water gas

Gas obtained by passing steam through a bed of coal or **coke**. The reaction taking place is:

$$C + H_2O \rightarrow CO + H_2$$

Gas so produced has a **calorific value** of 10–11 MJ m^{-3} and is called blue water gas on account of its flame colour. If the end product is to be a gaseous fuel for reticulation, the blue water gas will usually be raised in calorific value by conversion to **carburetted water gas**. Otherwise, the CO/H$_2$ blend resulting from steam gasification of the coal might be used as **synthesis gas**.

BOC GH2OST

Name assigned to a vehicle developed with backing by a group of organisations including British Oxygen. It is powered by a **fuel cell,** which has a higher intrinsic efficiency than processes involving thermodynamic cycles. The vehicle is constructed of lightweight material and has seating space for the driver only.

Boghead coal

Non-banded, translucent high-volatile **bituminous** or **sub-bituminous** coals derived from algal residues. Similar in appearance and combustion properties to **cannel coal**s.

Bohai Sea, China

The scene of formation of an oil slick about 2.5 miles long and 1.5

miles wide when a Chinese ship collided with a berthed oil tanker in 2002. The tanker was registered in Malta.

There has been a catalogue of such incidents in **Chinese waters**. For example, in November 2004, 450 tonnes of oil were released into the Pearl River in south China as a result of the collision of two vessels and the slick resulting was 17 km long. In May 2001 a vessel carrying styrene leaked 700 tonnes of its payload in the waters off Shanghai: the styrene-bearing vessel was from South Korea and the vessel with which it collided from Hong Kong. In November 2000, 230 m³ of oil leaked into the sea adjacent to Hong Kong in a collision involving an oil freighter. In October 1994 a tanker owned and operated by the Huahai Company of Beijing released its contents and polluted beaches and reefs.

Boiler

A closed vessel in which water is heated to generate steam, which may be superheated under pressure or vacuum through the application of heat, following combustion of a fuel or from the recovery and conversion of unused energy. There are two types of boiler. In the first, heat is applied to tubes containing water and steam (e.g. in a coal-fired power station boiler). In the second, combustion gases pass through tubes surrounded by water and steam (e.g. in a steam train boiler).

Boiling liquid expanding vapour explosion (BLEVE)

A type of explosion that can occur when a vessel containing a pressurised liquid is ruptured. When the liquid is water, the explosion is usually called a steam explosion. A BLEVE does not require a flammable substance so is not usually considered a type of chemical explosion. However, if the substance involved is flammable, it is likely that the resulting cloud of the substance will ignite after the BLEVE, forming a fireball and possibly a fuel-air explosion. BLEVEs can also be caused by an external fire near a storage vessel, heating the contents and causing pressure to rise.

Bolivia, energy scene in

This South American country has approximately 450 million **bbl** of recoverable crude oil and this is expected to increase as exploration activity by foreign oil companies continues. Three quarters of the oil is in the southwest of the country. There are non-associated gas fields containing a total quantity of about a million m³: this excludes the newly discovered field at **Ipati**, which is being developed by Total. Bolivia is a net exporter of electricity, generating ≈1150 MW during 2002 and utilising domestically about 90% of that. There is very little coal in Bolivia.

Bomb calorimeter

Device whereby a fuel is burnt under pure oxygen in a steel vessel (bomb) and the **calorific value** determined from temperature measurements made during the operation. Because the burning is at a constant volume, the calorific value equals the change in **internal energy** whereas in a **flow calorimeter**, which operates at constant pressure, the calorific value measured is the **enthalpy** change.

Bombay High Field

Offshore oil field off western India, about 100 miles from Mumbai (Bombay) and the scene of the loss of ten lives in July 2005 when a platform, operated by the state owned Oil and Natural Gas Corporation (ONGC), was destroyed. The accident appears to have occurred when a mobile drilling platform collided with a production platform. The production platform which was destroyed had produced 100 000 **bbl** per day, a seventh of India's domestic oil. The entire field produces about two and a half times that. Rescue operations were by helicopter (those of the ONGC and those of the Air Force), lifeboat and ship.

Boney

Term used in the eastern US for the waste from **bituminous coal**, the approximate analogue of **culm** for **anthracite**. It finds application as fuel for power generation at places including **Bakerton, PA**.

Bosa Chica

Nature reserve in southern California which, in 1990, suffered significant pollution when the tanker American Trader released about 7000 **bbl** of crude oil in nearby waters.

> This came just over two months after **Exxon Valdez**. Other spillages into US waters in recent years have included the 1996 incident in Galveston Bay when 100 bbl were leaked from a **barge**. In July 2000 there was spoilage of one of the most productive fishing areas off Rhode Island when a

> barge collided with the tug that was towing it. Later that year the Florida coast was impacted by release of crude oil from a tanker which had not collided with anything but simply discharged part of its cargo into the sea.

Botswana, energy scene in

Like all other southern African countries except Angola, Botswana has no significant domestic oil and gas. It is however very rich in coal, there being > 7000 Mt of provable reserves though production is barely 1 Mt yr^{-1}. This is produced at the mine at Morupule and is used in power generation at about 120 MW, roughly half the electricity consumption. Botswana accordingly imports the remainder of its electrical power from South Africa.

Bottom ash

The incombustible inorganic matter residue following coal combustion that has fallen to the bottom of the **boiler**, too heavy to be entrained in the gases as **fly ash**. Bottom ash is a marketable product used, for example, in making breeze blocks.

Bowen Basin coalfields

In Queensland, Australia, produces **bituminous coal** suitable for coke manufacture. Expected production figures for 2007 are 2 million tonne, some of it destined for the export market as coking coal. Total recoverable reserves are believed to be just under 100 Mt. There is joint activity with certain mining companies in the provision of coke for production of steel.

Boyce, Louisiana

The scene of a projected power facility using solid fuels in a **fluidised bed**. As well as coal, **petroleum coke** from the local refining industry will be used as fuel, generation will be at 600 MW and start-up will be in 2009. The operator is the Cleco Corporation who supply over a quarter of a million customers in Louisiana with electricity, having in total a generating capacity of 1.3 GW.

BP statistical review of world energy

The Review is published by BP each June and tells the story and history of world energy through the numbers behind the energy market headlines. It is an invaluable source of information for anyone wishing to track energy reserves and production over time. The Review is free to access, including spreadsheets and presentations, via BP's website: http://www.bp.com.

BPSD
See **Barrels per stream day.**

Braemar field
Condensate field in the North Sea 200 miles from Aberdeen. The field is in shallow water (about 125 m) and is believed to contain 3 bcm of gas and 10 million **bbl** of condensate. Production at the field began in 2003 and participants include Marathon.

Brash
A term used in the forestry industry, approximately equivalent to **forest thinnings** though having a somewhat different nuance. It is sometimes a requirement in forest management that some brash is left on the forest floor so that its decomposition will provide nutrients, therefore only a specified proportion is released for fuel use. A forest floor comprises both brash, which has been deliberately cut away, and dead material having dropped from the trees. In forestry this is called 'forest litter' and its density (weight per hectare) at any one time is an indicator of forest fire risk. When litter is actually purchased as a fuel it is simply called 'forest residue'. 'Baled brash' is one form in which waste wood fuel is sold to users for applications including **CHP.**

Brassica rapa
Botanical name for the plant from which **rapeseed** is obtained.

die Braunkohle
The German word for **brown coal**, of interest because a great deal of brown coal technology was developed in Germany in the late 19th and early 20th centuries. Such technology was later applied in the US and in Australia.

der Braunkohlenkoks literally means 'brown coal coke', although **brown coals** do not form a **coke** but a **char**, the distinction being lost in the translation from German. More importantly, a disadvantage of brown coals is that their solid **pyrolysis** residue is not suitable for use as a metallurgical reductant simply because being a powdery material it cannot support the payload of metal ore which a suitable coke, having good mechanical strength, can. There have however been 'hard chars' produced from brown coals and these are interchangeable with coke for metallurgical use. Certain brown coals will yield a hard product on pyrolysis if heating is not external as in a retort but internal by passage hot gas into the bed of coal. The gas enters the voids between the particles of coal and after pyrolysis a 'hard char' remains. Such chars have been made on a large scale in Australia with coal from the **Latrobe Valley**, and these have been widely exported.

Brayton cycle
A thermodynamic cycle, the basis of gas turbine operation. A Brayton cycle has significant conceptual similarity to the **Rankine cycle**, there being two isentropic and two isobaric steps in both cycles. However, whereas the Rankine cycle uses a condensing fluid (steam) as the working substance, the Brayton cycle uses air, heating being provided not by phase change but by use of a suitable fuel.

Brent Bravo
Oil and gas platform in the **Brent field**, the scene of two accidents close together in time, one of them fatal. As noted in the entry below, this platform has concrete supports. Two of them are filled with seawater and another—the utility leg—contains stores and equipment. In September 2003 two men died from suffocation through a gas leak while in the utility leg. The company operating the platform were heavily fined. In August 2005 there was a minor oil leak at the same platform. Twenty-six of the eighty-six persons at the platform were evacuated to a nearby installation. There were no deaths or injuries. The platform was shut down while the leak was repaired. Its output is 10 000 **bbl** of oil per day.

Brent field
Oil field in the North Sea that has been productive since 1976. There are a family of four production platforms: Brent Alpha, **Brent Bravo**, Brent Charlie and Brent Delta. Brent field oil when blended is an important **benchmark crude**. Three of the four platforms at Brent have **jacket** support while one—Brent Bravo—has concrete supports, having been built at a shipyard in Norway and towed to the site of the three concrete supports.

Brent Spar
Oil storage facility in the North Sea, operated by **Shell**, the decommissioning of which in the mid 90s was the source of some

controversy. Shell's intention was to sink it, a plan which had government approval, but there was active protest by Greenpeace. Brent Spar was eventually towed to Norway and recycled in the construction of a new ferry terminal.

Bretforton, Worcestershire

The scene of permanent illumination of bus shelters and other street locations using **photovoltaic cells**. Lighting is by fluorescent tubes each rated at 9 W. A total of 150 such installations are in use, and it is expected that the use of photovoltaic cells for flashing lamps at school crossing patrols will follow.

Briquettes

Solid fuel made from small particles compressed and moulded to a regular shape such as a spheroid or a cuboid. Almost all solid fuels have been so processed. **Bituminous coal** fines can be briquetted to give a good general-purpose solid fuel. **Lignites** are extensively briquetted the world over using technology developed in 19th Century Germany. Waste from the processing of cotton, flax and other natural fibres can be made into biobriquettes.

Generally briquetting will require a binder. Starch solution and petroleum refining residue are common binding agents. Some lignites do not require a binding agent for briquetting as the minerals within the coal, especially clay, fulfil that role. Briquettes made in this way have a limited storage life before they start to disintegrate, especially if kept out of doors. Briquettes have been used not only for burning but also to make fuel gas, for reticulation to homes and commercial premises, by reaction with steam.

Britannia field

Non-associated gas field in the North Sea, also producing NGL/condensate, the daily yield being typically 23 000 **bbl**.

British Columbia, wind power in

The BC provincial government has offered suitable land rent free to organisations able to establish wind farms on such land. Rent will be payable from the beginning of the 11th year after generation begins. The largest **wind farm** to be built under this scheme is the Naj Kun wind farm which has a coastal location and will, when complete, have 700 **wind turbines**.

British Thermal Unit (BTU)

The Imperial unit of quantity of heat still widely used, notably in the US. Its initial replacement was by the calorie which conforms to the mks (metre-kilogram-second) system of units but is not SI. The SI unit for quantity of heat is the **joule**. Conversion requires the following factors:

$$1 \text{ BTU} = 252 \text{ calories}$$

$$1 \text{ calorie} = 4.2 \text{ J}$$

The unit *therm* once widely used in fuel gas pricing is 1000 BTU.

British Thermal Unit (BTU) conversion

Term applied to methods by which **coal** can be utilised in a cleaner way than by its direct combustion. Such methods include coal **gasification** and the generation of electricity from coal.

Brown coal

The least mature **coal** along the **coalification** sequence. There are huge deposits in many countries including Germany, the US (Montana, N. Dakota and Texas) and Australia. Such coals are soft when in the bed moist state and in an open cut can be mined by dredging rather than drilling. Newly mined they have a **calorific value** of 8–10 MJ kg^{-1} rising to > 20 MJ kg^{-1}, depending on factors including the ash content, on drying. Within a single deposit the colour will vary visibly from pale brown to very dark as a result of different **maceral** compositions. Reactivity towards oxygen declines as coalification takes its course and brown coals are relatively very reactive.

Brown coals have found extensive use, for example in power generation. The brown coals of the Latrobe Valley in southeastern Australia have, for eighty years, provided most of the electricity for Melbourne (current population > 3 million). For power generation they are burnt as **pulverised fuel** (pf) having been milled to particle sizes in the range10–100 μm and admitted with air to a furnace to give a jet flame, the post-combustion gas containing **fly ash**. Brown coals have also been the basis of huge quantities of **briquettes**.

The high intrinsic reactivity of brown coals makes them susceptible to **spontaneous combustion** in storage and necessitates measures to prevent it. They are also more porous than higher rank coals. Entry of water vapour into the pores can lead to a 'heat of wetting' effect which exacerbates spontaneous heating. Brown coals are very responsive to solvents including **acetone** and **methanol**, interaction of a coal with a solvent being manifest as swelling to the extent of a 30–40% increase or more in their initial bulk volume. The brown coals of the world vary widely in their ash-forming tendencies. Those originating from Thailand are particularly high in ash.

World brown coal production for 2003 was 886 Mt. The largest single producer was Germany which contributed 180 Mt to the world total.

Brown's Ferry

Nuclear power generation facility in northern Alabama, owned and operated by the **TVA** and operational since 1977. Using U^{235} as fuel the facility is capable of generating at 2285 MW. It comprises three reactors, one of which has been out of service since 1985 although the TVA Board recently voted to put it back into service.

Bruce Field

Non-associated gas field in the North Sea, discovered in 1974 and productive since 1993. Reserves are 2.6 trillion m^3 of gas and 250 million **bbl** of condensate. Current production is 17 million m^3 per day of gas and 80 million bbl per day of condensate. The **jacket** of the production platform at Bruce field has recently been fitted with a new riser to take gas piped from the **Rhum field** 25 miles away.

BTX

Acronym for Benzene (C_6H_6), Toluene ($C_6H_5CH_3$), Xylenes ($C_6H_4(CH_3)_2$) which often occur together for example in the manufacture of **coke**. Because of their similar molecular structures, these compounds are very difficult to separate so require the tallest fractionating column in a **refinery**. BTX is toxic and exposure limits apply. BTX can be the starting material for the manufacture of octane enhancers. There are three different isomers of xylene: ortho (*o*-), meta (*m*-) and para (*p*-).

benzene

toluene

o-xylene

m-xylene

p-xylene

Bu Kadra Bridge, Dubai

The scene of a tanker explosion which was carrying oil by road, necessitating major repairs. Since the accident oil tankers have been prohibited not only from the bridge but also from road tunnels.

Bubble point

The pressure at which the first infinitesimally small amount of vapour appears in an all liquid system at the system temperature. See **dew point**.

Bukom refinery

Situated in Singapore, at the present time the largest of all the refineries in the world operated by Shell. It processes 500 000 **bbl** per day and a significant proportion of its products are exported to other Far East nations. Shell are active in petrochemicals on **Jurong Island**. Shell's involvement in the **Sakhalin Energy** project is co-ordinated from Singapore.

Bulgaria, hydrocarbon reserves of

There is significant onshore oil and gas production at fields operated by Pleven Oil and Gas. There is currently major exploration activity about 50 miles from the capital Sofia.

Bulk density

The density of a powdered substance such as **coal** or a fibrous one such as **bagasse** calculated according to:

$$\text{bulk density} = \frac{\text{mass of the substance}}{\text{volume of container}}$$

It is clearly lower than either the **mercury density** or the **helium density** of the substance and, unlike either of those, has a dependence on the particle size. The substance to which the bulk density applies is of course actually two-component: the solid particles and the air in the voids. Other properties including the thermal conductivity are for the two-component system and one would not apply the thermal conductivity for a single coal particle to an assembly of such particles with air between them.

Bulk density also relates to **refractory materials**, both the final product and the crushed or powdered solid starting material.

Bulk terminal

A storage facility that can receive **petroleum** products by barge, pipeline or tanker, until required for shipment to a **refinery** or other user.

Buncefield oil depot

A major facility for storage and distribution of hydrocarbons. It receives refined products from places including Canvey Island in the Thames estuary. Gasoline and fuel oil from Canvey Island are transferred by pipeline to Buncefield. Another source from which the Buncefield terminal receives fuel for distribution is the Fina refinery in Lindsey, Lincolnshire. Buncefield is in the business of distribution, and hydrocarbons which pass through the facility are already refined and ready for their intended uses. As well as gasoline and heating oil, aviation fuel is received at Buncefield for transfer by pipeline to the airports of London.

A serious accident occurred there early on Sunday 11th December 2005. There were in excess of 30 non-fatal injuries. It was several days before the flames were extinguished.

Bunker fuel

Term used in shipping for the fuel—coal or fuel oil—which powers a vessel. The term is sometimes broadened to include these fuels in applications other than marine.

Burgan field, Kuwait

Oil field having a production of 1 million **bbl** per day and often considered to be the second largest oil field in the world, after the **Ghawar field**, Saudi Arabia. The Burgan field has been yielding oil since 1938 and its estimated reserves are 70 billion bbl. Discoveries of oil continue in Kuwait, there having been a major find as recently as 2003. This is now called the Kara al-Marou field and contains 1 billion bbl.

Burner

Device for stabilising a flame. A flame is intrinsically a dynamic phenomenon, propagation always being towards unburnt fuel. In a burner, one of the simplest examples of which is the Bunsen burner, the burning velocity and the speed of influx of the cold gas/air mixture are equal and opposite, causing the flame to be stationary. If there is a small surge in the gas supply, raising the flow speed of the influx gas, the flame moves upwards slightly (although probably not to an extent visible to the naked eye) so as to lose less heat to the metal mouth of the burner. It therefore adjusts its own speed to match that of the cold gas so that mechanical stability is maintained. Conversely, if the influx gas drops in velocity, the flame moves a little closer to the burner mouth, losing more heat to it and adjusting its speed downwards so as to match that of the cold gas. Any one burner can affect such adjustment only across a limited range of conditions of thermal supply, that is, rate of supply of fuel/air to the burner. Also relevant to burner performance and safety are the reactivity of the gaseous fuel proposed for use and its calorific value. A gas will burn safely only on a correctly designed and adjusted burner. To attempt to burn, for example, **liquefied petroleum gas (LPG)** on a burner designed and adjusted for natural gas is highly dangerous unless the LPG has been modified to **tempered LPG**.

A burner using a liquid fuel is likely to need high levels of electrical power for two purposes: to pump the fuel (which, if it is a petroleum residue, is likely to be quite viscous) and also to pre-heat it to generate sufficient vapour for ignition. There have been many cases of electrocution during operation and maintenance of such burners. Liquid entering a burner is dispersed into droplets by the process known as **atomisation** and depending on how rapidly it evaporates will burn either similarly to a gaseous flame (almost immediate evaporation) or similarly to a flame composed of fine coal particles in air (sluggish evaporation).

There are burners which are dual-fuel, receiving for example natural gas and fuel oil simultaneously.

Burnup

Term used for the rate of heat release by a nuclear reactor, most common unit GW day per **tHM**. Values for reactors in power generation are expected to be tens of GW day per tHM.

Burton, William Merriam (1865–1954)

Developer of thermal **cracking** of crude petroleum products. The process which he patented in 1915 was directed at breaking down fractions of crude oil heavier than gasoline in order to bring them within the gasoline carbon number and boiling ranges so that they could be blended with the gasoline to 'extend' it. His process was in wide use for 15 years during which it saved an estimated 1 billion **bbl** of crude oil. In the 1930s new cracking technologies, usually involving a catalyst, were developed to make olefins for the then emergent petrochemical industry. These were also applied to the breakdown of heavier petroleum material to extend gasoline, and Burton's process became less widely used.

Burundi, energy scene in

Like many other countries in the same region, Burundi has hydroelectric resources. These produce energy at about 40 MW, and a quantity of electricity equivalent to about a quarter of this is imported into Burundi. The low total electricity reflects the fact that less than 2% of the population of this country of 6.4 million has access to electricity. There is significant usage of **biomass**, **charcoal** and **peat**.

Busbar

The beginning of the electrical transmission system at the power plant. The cost of producing 1 kWhr of electricity and delivering it to the transmission system is known as the 'busbar cost'.

Butane

A component of natural gas, up to 20%, butane (C_4H_{10}) is used in **liquefied petroleum gas (LPG)** and **natural gas liquids (NGLs)**. It is also a product of **cracking** petroleum. Butane is highly flammable and colourless. The two isomers of butane are normal (n-) and tertiary (t-).

n-butane

t-butane

Butylene

A highly flammable and easily liquefiable gas, butylene (or butene) has three isomers (C_4H_8). Butylenes are used for making synthetic rubber.

but-1-ene

but-2-ene

iso-butene

Buzzard field

Newly developed oil field in the North Sea, where operators include BP and British Gas. Discovered in 2001, Buzzard is seen as being one of the most important finds in the UK sector of the North Sea in recent years. Oil was first produced there in January 2007, and production by mid 2007 is expected to be 200 000 **bbl** per day, with **associated gas** in amounts of 1.7 million m^3 per day. The field is in the Firth of Moray, and oil from it is taken to a terminal in S.E. Scotland.

C

Caister field

Non-associated gas field in the North Sea of the Lincolnshire coast. Operated by Total, it commenced production in 1993. It shares some infrastructure with the neighbouring Murdoch field. Both have reached the stage where internal well pressures are low therefore a compressor is needed, this having been manufactured by Rolls Royce.

Caking

A term used to describe a coal that fuses to a solidified mass on heating. Two methods are used to determine the caking power of a coal: the Roga index and the Gray-King assay.

The Roga Index: A mixture of known masses of crushed coal and a standard **anthracite** are heated rapidly in a crucible to 850°C. The resulting coke is examined in a Roga drum for its resistance to abrasion and, hence, the Roga index is calculated. The index reflects variations in rank, petrographic composition, mineral matter content, and whether the sample is weathered or oxidised. It is not dependent on pressure.

The Gray-King assay: The coal is heated slowly at 5°C min^{-1} to about 600°C and the resulting **coke** examined visually and graded from A to G. For strongly caking cokes, greater than type G, the number of parts of electrode carbon x required to be mixed with 20–x parts by weight of the sample to obtain a coke of type G are calculated. G_9, for example, indicates that nine parts of electrode carbon need to be added to the coal mixture to produce a grade G coke. The test is time consuming but the results reflect variations in rank, petrographic composition, mineral matter content, and whether the sample is weathered or oxidised.

Grade	Description
A	Pulverulent
B	Just coherent; breaks into powder on handling
C	Coherent but friable on rubbing
D	Shrunken, moderately hard
E	Shrunken and fissured, hard
F	Slightly shrunken, hard
G	Hard, no shrinking; standard coke
G_1	Slightly swollen, hard
G_2	Moderately swollen, hard
G_3	Highly swollen
G_9	Very strong swelling, increased length

Calder Hall, UK

Scene of the first commercial nuclear power plant in the world, which commenced production in 1956. It had a maximum generating capacity of just under 200 MW and remained in operation for almost 50 years. Decommissioning is now underway.

Calder Hall was followed by the Chapelcross nuclear power station in southwest Scotland in 1959 and this is in fact still in operation, again at close to the 200 MW level. Not very far from Chapelcross, a little south of Edinburgh, is the Torness power station, a much more recent installation which is expected to remain in service until 2023. It has a capacity of 1250 MW.

Calorific value

The quantity of heat released by a unit amount of a particular fuel. There is a dependence on conditions: in constant-volume burning the calorific value is the internal energy change (usual symbol ΔU) whereas in constant-pressure burning the calorific value is the enthalpy change (usual symbol ΔH). However, routine calorimetry can seldom distinguish the two and it is fairly immaterial whether we consider the calorific value to be ΔU or ΔH.

The calorific value is usually on a weight basis, for example a **bituminous coal** will have a calorific value in the neighbourhood of 30 MJ kg^{-1}. For gases, unit amount is 1 m^3 at 288 K, 1 bar pressure. On this basis, natural gas has a calorific value of about 37 MJ m^{-3}.

Note that the term 'calorific value' is still appropriate after the abandonment of the calorie as the unit of energy. To express a calorific value in joules per unit amount is quite acceptable. Synonyms for calorific value are 'heat value' or 'heating value'. The latter is the basis of two acronyms: GHV (gross heating value) and NHV (net heating value). When a fuel burns, the GHV is obtained if water formed as a product condenses and the latent heat so released adds to the heat of combustion. The NHV is obtained if the water remains in the vapour phase. GHV and NHV are sometimes referred to as HHV (higher heat value) and LHV (lower heat value). Clearly, a fuel with negligible hydrogen, perhaps a **char** prepared at high temperature, would have the same values for the HHV and the LHV.

Caltex, refining activities in the Far East

The interests of this company in the Far East include the massive Yosu refinery in South Korea, with a capacity of 650 000 **bbl** per day, which also has **cracking** facilities on site. This refinery is actually the third largest in the world, its products required to sustain South Korea's vibrant and expanding manufacturing sector. Almost all of the crude that this facility receives is imported, most of it from Dubai. Caltex also has a significant stake in the **Rayong refinery** in southern Thailand and in a refinery in Pakistan.

Calypso US pipeline

Projected pipeline linking the Bahamas with Florida. **Liquefied natural gas** taken by ship to the Bahamas will be converted back to gas by heat exchange, and piped 90 miles to Broward County, Florida, where it will be used to raise steam for power generation. The pipeline is planned in response to rises in electricity consumption in Florida and will have a capacity of 25–30 million m³ of natural gas per day potentially capable of generating about 3000 MW of power.

There are also plans to build a refinery in the Bahamas which will take Iranian crude. Heavy distillate and/or residual fuel oil produced will be used to make electricity, some of which will be sold to Florida.

Cambodia, energy scene in

Wood is the staple domestic fuel. In fact, Cambodia has the best forest assets of the entire mainland Asia even though their management in recent years has been lamentable: it is believed that illegal exports of timber, largely to Thailand, exceed legal exports.

Cambodia uses electricity at about 80 MW, 35% of it **hydroelectricity**, the remainder generated with imported oil.

Cameroon, oil production by

This country has significant offshore oil fields. Production began in 1978 with Elf as one of the participants. Current production is about 35 million **bbl** per year, more than half of which is exported to the US or Europe.

Campomento, Spain

Scene of a new **liquefied natural gas (LNG)** terminal, which will receive shipments from **Qatar** every three days. The LNG arriving at the terminal will, after conversion back to gas, be transferred by pipeline to Italy. Campomento has been functioning as a port for 100 years.

Canadian Arctic, natural gas reserves of

There are known to be 2.5×10^{11} m³ of recoverable **natural gas** in this region of the world and the **Mackenzie pipeline** is proposed as one way to take it to population centres. Other possibilities are the production of **gas-to-liquid fuels** or **liquefied natural gas**.

Candle

Consisting of a column of solid fuel surrounding a wick, the candle is a source of light and heat. Before being lit, the wick is saturated with the fuel (usually paraffin wax) in its solid form. The heat of the ignition source first melts and then vaporises a small amount of the fuel. This then combines with oxygen in the atmosphere to produce a flame. Heat from the flame melts the top of the solid fuel; the molten fuel then moves upward through the wick via capillary action. Finally, the fuel is vaporised to burn within the candle's flame.

The burning of the fuel takes place in distinct regions. Within the bluer, hotter region, hydrogen is separated from the fuel and burned to form water vapour. The brighter, yellow portion of the flame indicates that carbon is being oxidised to form carbon dioxide. The earliest known candles were around in 3000 BC when the Egyptians used beeswax as the fuel.

> The most famous scientific discourse on the candle was given by Michael Faraday (1791–1867). He gave a series of six lectures on the "Chemical History of a Candle" at the Royal Institution in London, first published in 1860. This was the first in the series of *Christmas Lectures for Young People*, which still continues to this day. Faraday's discourse began:
> "There is no better, there is no more open door by which you can enter into the study of natural philosophy than by considering the physical phenomena of a candle."

Cannel coal

Coal of dark grey to black colour lacking the band structure that coal deposits usually have. Such coals are rich in volatiles and have been seen as promising starting materials for the production of liquid fuels from coal by **pyrolysis**. The nature of the starting vegetation in the formation of such coals differs from that of other coals and is sometimes believed to have been seaweed rather than trees, shrubs or mosses on the Earth's surface.

Canneloid

Cannel coal having attained a high degree of **coalification**, the analogue of **anthracite** for conventional coals.

Canola seed

Synonym, widely used in the US, for **rapeseed**.

Canunda, South Australia

Parkland region and site of the largest wind farm in the southern hemisphere. It has 23 turbines each with a 2 MW capacity, and came into full operation in January 2005.

Canyon Reef oil field

Oil field in Texas, currently producing about 20,000 **bbl** per day, which was in the business of carbon dioxide storage 25 years before the **Kyoto Protocol**. Since 1972 it has been the practice at the oilfield to inject carbon dioxide into the wells to raise the crude oil production. There are 2000 miles of pipeline bringing carbon dioxide to the Canyon Reef field. Some of it has been separated from hydrocarbon gases using **monoethanolamine**, while the remainder is from one of the several **carbon dioxide fields** in that part of the US.

Clearly this enterprise was ahead of its time with the result that in the post-Kyoto years its collaboration has been widely sought.

Cape Cod wind project

Proposed construction of a **wind farm** in Nantucket Sound, MA. Predictably, opposition is fierce and the project is at the time of going to press a long way from being finalised. The plans for the wind farm include 130 **wind turbines** each with a capacity of 3.6 MW giving a nameplate capacity for the wind farm of 468 MW and an actual production of about 170 MW. The local area, which includes Martha's Vineyard, has a requirement of 230 MW, about 75% of which could therefore be supplied by the wind farm.

CAPP

See **Central APPalachian.**

Capping

The process of closing, or a device used to close, an oil or gas well to prevent escape.

Car manufacture, electricity consumption by

As an example we consider the Nissan plant in Tyne and Wear, UK, one of the largest car manufacturing plants in Europe having a payroll of 4000. It consumes electricity at ≈ 500 MW. The company is planning to install **wind turbines** that will eventually meet 10% of its power requirements.

Carbon, elemental

The key to life, carbon is a very common element with symbol C. Its atomic number is 6, and atomic weight 12.01115. Carbon has the unusual ability to bond with itself chemically as well as with other elements so is known to form almost 10 million compounds. It is present in the atmosphere as carbon dioxide.

Carbon black

(1) Term sometimes used synonymously with smoke.

(2) Carbon particles yielded as a desired by-product of processes including the **cracking** of hydrocarbons to give carbon and hydrogen. One application is to the manufacture of tyres where they are added to the primary rubber constituent.

Carbon dioxide, concentration of in the atmosphere

Currently about 370 ppm. In the early 19th Century, at the onset of industrialisation, it was about 280 ppm. In New Testament times it was about 260 ppm. These data are known from 'ancient air' trapped in bubbles of known age in the Antarctic ice layer. Factors responsible for the rise are fuel utilisation and deforestation. A rise of 90 ppm over two hundred years converts to 0.45 ppm per year but this is not in fact the case: the rise is not linear and in the early 21st Century it exceeded 1 ppm per year. In weight terms, this translates to an additional 8000 Mt yr^{-1} of carbon dioxide in the atmosphere.

Carbon dioxide credits, penalty rates for

An organisation requiring more carbon credits than it has been allocated can buy them on a world market. Compliance with emission allowances is of course audited by environmental authorities in the various countries who will themselves be in possession of credits and transfer them compulsorily to an organisation 'in debt' at a price two to three times the market price.

Carbon dioxide, predictions for 22nd century levels

Industrialisation of the world raised the CO_2 concentration of the atmosphere by about 100 ppm to ≈ 370 ppm. Not only is the level rising by about 1 ppm per year at present, but the first derivative of that level with respect to time is also rising because of increased usage of fossil fuels. Eventually, it is hoped, a point will be reached where measures including the use of carbon neutral fuels, together with increases in **carbon sequestration**, will achieve the condition where CO_2 levels are constant (at about 500 to 550 ppm). It is anticipated that this point will be reached by the year 2100.

Carbon dioxide, production of in thermal generation of electricity

About 550 g of CO_2 per kWh is produced when natural gas is used as fuel in a steam turbine operating on a Rankine cycle with 33% efficiency. Other things being equal, this is roughly half the amount of

CO_2 produced compared to coal of 80% carbon content with a **calorific value** of 30 MJ kg^{-1}. That explains the swing to natural gas for electricity generation in some parts of the world where coal is available and has previously been used for many years to raise power.

Carbon dioxide, solubility of in coals

The consensus among those specialising in the physical chemistry of coals is that carbon dioxide dissolves significantly in any coal. Of course, the pressure of carbon dioxide gas in phase equilibrium with carbon dioxide so dissolved needs to be very high. Pure carbon dioxide at ordinary temperatures and 1 bar pressure is a long way below the pressure at which it is in phase equilibrium with the liquid form. Pressures of this order are required for significant dissolution of the compound in coal. This is an issue of importance when redundant coalmines are used for **carbon sequestration**, where the pressures of carbon dioxide *are* sufficient for dissolution. The CO_2 dissolution, also referred to as imbibition, causes coal swelling which changes the volume of the space into which the carbon dioxide is injected. Recently, this has been experienced as a difficulty in underground carbon dioxide storage at disused mines in Poland.

Carbon dioxide fields

Natural reserves of carbon dioxide, including those in Utah, Oklahoma and Colorado, which are owned and operated by oil companies such as Exxon Mobil and BP. The McElmo Dome field in Utah provides some of the carbon dioxide used at the **Canyon Reef oil field** for well injection. The carbon dioxide from this field is 98% pure and its estimated reserves are 3×10^{11} m^3, equivalent to 5×10^8 tonne. The carbon dioxide fields of the southwest USA provide collectively ≈ 20 Mt per year of the gas for use at oil fields.

Carbon monoxide

Chemical compound, formula CO. It is flammable, pure carbon monoxide having a calorific value of about 11 MJ m^{-3}. It is also fatal to human beings if inhaled to a sufficient degree. A concentration of 5000 ppm ($\equiv 0.5\%$) will cause loss of consciousness in about 5 minutes.

The importance of carbon monoxide in fuel technology is threefold:

1. It is a constituent of many manufactured gases including **producer gas**, **retort coal gas** and coke oven gas.

2. It is present in post-combustion gas, where it is a toxic hazard.
3. It is a **devolatilisation** product of coals.

Ambient levels of carbon monoxide are usually a few tens of ppm.

The **calorific value** of carbon monoxide on a molar or volume basis is only 1% different from that of elemental hydrogen. The equality of molar weights means that if nitrogen is also present, amounts of CO cannot be measured by mass spectrometry. The equality of calorific values means that in calculating the calorific value of a gas containing both hydrogen and carbon monoxide, they can be grouped together. Note that in **producer gas**, hydrogen and carbon monoxide are likely to be the only two combustible gases present.

Combustion of the compound in pure oxygen in laboratory studies has revealed quite unique behaviour. There is a pale blue emission which might be steady, lasting up to about 20 seconds, or oscillatory. The intermediate responsible for the glow is electronically excited carbon dioxide. There has been much debate as to whether the glow can occur under conditions totally free from hydrogen atoms.

Carbon neutrality

Term applied to the burning of wood when its contribution to the **greenhouse effect** is considered. The carbon content of any carbon dioxide molecule released by burning wood was previously present in a carbon dioxide molecule in the atmosphere before being stored by the tree, hence the carbon neutrality. The residence time of a carbon dioxide molecule in the atmosphere, before being removed and stored by e.g. forests, plants and oceans, is of the order of tens of years.

Carbon sequestration

The removal of carbon dioxide from post-combustion gas, in order to prevent its admittance to the atmosphere where it would contribute to the greenhouse effect and global warming. Carbon dioxide so removed can be converted to a liquid having a density just below that of liquid water and stored as such in containers which can be buried. There is research interest in the disposal of liquid carbon dioxide in such places as disused coalmines or at depleted offshore oil and gas fields, for example at the **Sleipner field**. As an illustration of the scale of the storage space required, a power station using **anthracite** and generating power at 2000 MW with an efficiency of 35%, would produce sufficient liquid carbon dioxide to fill a cube of side 35–40 m in one day's continuous operation. There is also activity in places including **Weyburn, Saskatchewan** in disposing of gaseous carbon dioxide by injecting it into oil wells in order to effect **enhanced oil recovery**.

Carburetion

Incorporation into a thermodynamic cycle which uses air as the working substance, such as the Otto or diesel cycle, of something capable of chemical heat release. A hydrocarbon vapour is the usual choice.

Carburetted water gas

Gas with a **calorific value** of about 20 MJ m^{-3}, the same as that of retort coal gas, although the two differ significantly in composition. Carburetted water gas is made by adding hydrocarbon gases, made by cracking oil-based material, to **blue water gas**. It is an example of **town gas**.

Carlsbad, NM

Scene of an accident due to natural gas pipeline failure in 2000, killing 20 people who were camping nearby. The previous year at Bellingham WA a pipeline carrying refined petroleum material leaked and formed a fireball; three young people were killed. The Carlsbad and Bellingham accidents, so close together in time and both attributable to pipeline failure, led to the development of new industry standards for pipeline safety.

Carson, CA

Scene of a refinery in southern California operated by ConocoPhillips in co-ordination with another, owned by the same company, a few miles away at Wilmington. The capacity of the combined facility is 139 thousand **bbl** of crude per day, and crude is received from within California. Refined products are distributed in southern California, Nevada and Arizona. Similarly, in northern California, ConocoPhillips operate two refineries together: one is in San Francisco Bay and the other in Santa Maria. They are linked by a pipeline and have a combined capacity of 106 thousand bbl per day.

Casing

A heavy steel pipe or tubing lowered into a borehole and cemented into place to prevent oil, gas or rocks from entering the hole. Once a well is ready to be exploited, explosives are used to make a hole in the casing at the appropriate point to allow for the oil or gas to be extracted.

Caspian Sea

The scene of huge amounts of offshore oil and gas, there being potential for further exploration on top of reserves which are already proven. Current development is intense and the target oil production from the Caspian by 2010 is 4 million **bbl** per day. Annual natural gas yields of $\approx 6.5 \times 10^{10}$ m^3 are expected by the end of the present decade. The **Baku-Tbilisi-Ceyhan Pipeline** has been installed to carry crude oil from a terminal at Baku on the Caspian coast via Tbilisi in Georgia to Ceyhan, a Turkish port on the Mediterranean.

The first discovery of **natural gas hydrates** was in the Caspian Sea in 1979 during **core** sampling there.

Catalytic cracking

The conversion of heavy oils into lighter, more valuable, products by breaking down the heavy hydrocarbon fraction of **petroleum** using heat (at approximately 650°C) in the presence of a zeolite catalyst. The process leaves the catalyst saturated in **coke**, which needs to be burned off in a regenerator before the catalyst can be reused. Often referred to as cat cracking; see **cracking**.

Catawba Power Station, SC

Nuclear power station in South Carolina with a capacity of 2258 MW. It entered service in 1985 and uses **mixed oxide (MOX)** as fuel. It is operated by Duke Power, as is the McGuire nuclear power station in NC which has about the same capacity. Both are expected to start using nuclear fuels produced at **Savannah, SC** in 2007.

Cavendish field

Newly developed gas field in the southern North Sea in **block** 43/19a. After processing at the field's own platform, gas is passed along to infrastructure also used for gas from the **Caister field** and from the Murdoch field. It is then taken to a terminal on the Lincolnshire coast. One of the participating companies in Cavendish field is the Aberdeen-based Dana Petroleum.

Ceiba field

Oil field just over 20 miles off the coast of Equatorial Guinea, acquired by Amerada Hess from Triton Energy in 2001 and now producing 26 000 **bbl** per day. Production is by means of a **floating production, storage and offloading (FPSO)**. The government of Equatorial Guinea have approved the development of adjacent fields by Amerada Hess. Fixed platforms are proposed for these, also a fifteen-mile pipeline which will take the oil to the FPSO at Ceiba for offloading onto tankers. Production from these fields is expected to begin in 2007.

Central APPalachian (CAPP)

CAPP coal is a benchmark in the pricing of coals. Specifications are as follows:

Calorific value	(28 ± 0.6) MJ kg^{-1}
Ash content	$< 13.5\%$
Sulphur content	$< 1\%$
Moisture content	$< 10\%$
Volatile matter	$> 30\%$
Hardgrove index	> 41
Unit amount	$= 1550$ tonne

(Particle size conditions also apply)

A coal conforming to all of these criteria will attract the daily price for a CAPP coal which fluctuates as trading takes place in the same way that the price of oil does. Such a coal is describable as high-volatile bituminous. CAPP price is for collection at the **Big Sandy River, WV**. Delivery charges are additional with the Big Sandy River as point of origin. The **Sago Mine, WV** is in the Appalachian region.

Centrifugal dewatering of coal

A means of removal of water after **coal cleaning**, and currently finding application at places including **Kalimantan**. There, 90% of the run-of-mine coal is export quality. The remaining 10% is washed and dewatered in a centrifuge before being blended with the other 90% for export.

Cepu field

Oil field in Indonesia not yet utilised in spite of having proven reserves of 500 million **bbl** of oil. Exploration and appraisal were by Exxon Mobil who formed the company Mobil Cepu which will eventually be involved in production from the field. Oil from the field will be taken to the **Tuban, East Java** refinery which will also process crudes from other Indonesian fields. The Cepu field will raise Indonesia's known crude oil reserves by 18%.

Cetane enhancer

Organic compound blended with **diesel**, for use in compression ignition engines, to improve performance. Currently widely used cetane enhancers include isopropyl nitrate (IPN) and ethyl hexyl nitrate (EHN). These substances, being nitrated organics (as is TNT) can react explosively without added oxygen and this has sometimes created difficulties in their storage and handling.

Ceyhan

Turkish port, terminus of the **Baku-Tbilisi-Ceyhan Pipeline** and also of an oil pipeline from **Iraq** capable of delivering 400 000 **bbl** per day. The latter recently came back into service, at about 60% of full capacity, after repairs following sabotage.

Ceylon Electricity Board

Although it has been over 35 years since Ceylon became Sri Lanka, the name of the electricity board has not changed. This country of 18 million inhabitants has no oil or gas and so there is a heavy reliance on wood fuel and other biomass including **bagasse**. Private firms generating electricity are encouraged to sell any surplus to the grid. **Combined heat and power** has been considered using bagasse from the sugar mills as fuel, although at the time of writing this has not come to fruition. The country imports a significant amount of fuel including all of its transport fuel.

Chain branching

Proliferation of reactive intermediates in a gas-phase reaction with accompanying acceleration of reaction rate. An example of chain branching is:

$$H + O_2 \rightarrow OH + O$$

where on the left there is one reactive intermediate and on the right there are two. Chain branching is important in understanding the mechanism of oxidation of hydrocarbons.

Char

Solid **pyrolysis** product of either a non-coking coal or of a combustible solid other than coal such as wood, **peat**, rubber waste, coconut shells, municipal solid waste. **Lignites** form chars, not cokes, when heated. Such chars tend to be powdery in nature though it is possible to prepare a hard char from certain lignites by making the pyrolysis process as even as possible by circulating hot gas in the charge of lignite instead of heating it externally. By far the dominant element in a char is carbon. Oxygen in the original chemical structure is removed, in the form both of oxygen-containing gases such as carbon dioxide and of oxygenated organic compounds in the liquid products, during pyrolysis. Consequently, chars are often saleable as starting materials for the manufacture of carbon materials including activated carbons.

Charcoal
A product of the thermal treatment of wood in the absence of enough air to prevent complete combustion from taking place, releasing volatiles. The product is a lightweight material composed of 70–90% carbon and a cleaner burning fuel than wood.

Chemicoil 1200
Trade name for a material which can be used to bridge, on a short term basis, breaks in oil pipelines. It is composed of polyester reinforced by polyurethane and is both flexible and resistant to chemical attack. Scenes of its use include the Druzhba pipeline that pumps oil from Russia to Europe via Belarus.

Chemiluminescence meter
A widely used method of measuring oxides of nitrogen in the atmosphere. It requires a cylinder of oxygen from which oxygen atoms, required for the reaction, are produced during operation.

Chequer
Brick-like material, uses of which include the packing of plant for the **carburetion** of **blue water gas**.

Chestnut coal
Anthracite having been prepared for use by crushing to the particle size range 3.02–4.13 cm.

Chestnut field
Oil field in the UK sector of the North Sea currently being developed. Its reserves are estimated to be 6–7 million **bbl** and production is expected to begin in the second half of 2007.

ChevronTexaco, lineage of
This corporation is active in 180 countries. Chevron and Texaco merged in the early 21st Century. Each had a long history, as did **Caltex** which is also part of the organisation.

Chevron's origins can be traced to 1879 when the Pacific Coast Oil Company was founded following the discovery of oil at **Pico Canyon**.

It was later acquired by **Standard Oil** and in 1926 was renamed Socal: Standard Oil Company of CALifornia. In 1936 Socal acquired 50% of the California Texas Oil Company, later Caltex, and in 1947 it acquired the previously independent Signal Oil Company. Meanwhile, at the opposite side of the US, the Gulf Corporation had come into being in 1907 and picked up some further assets along its way to merger with Socal to become Chevron in 1984. The Texas Oil Company was formed in 1902, a year after the discovery of the **Spindletop field**, and later renamed Texaco. It made a number of acquisitions including the Indian Refining Company (1931), and the Seaboard Oil Company (1958). In 1984 it made its greatest scoop by acquiring the Getty Oil Company. It merged with Chevron to form ChevronTexaco in 2001. In 2005 the name was changed simply to the 'Chevron Corporation' though trade use of the names Texaco and Caltex continues.

Chile, oil and gas reserves in

This South American nation has fairly depleted oil reserves, currently about 150 million **bbl** at onshore fields in the south. Chile consumes much more oil than she is able to produce and consequently imports large amounts from countries including Nigeria and Malaysia. Chile also receives crude oil by pipeline from two other South American countries: Argentina and Bolivia. There are three oil refineries in Chile the largest of which can receive over 100 000 bbl per day. Natural gas is also limited, there being about 10^{11} m^3 of recoverable reserves. Since 1997, Chile has been receiving natural gas from Argentina by pipeline from **Neuquen**. Argentina's own reserves of natural gas are not enormous (hence the involvement of the country in the **Ipati Block** in Bolivia) and supply to Chile will probably not be possible for many more years. Accordingly, a **liquefied natural gas** terminal in Chile is planned so that natural gas can be brought from more distant markets.

China, coal bed methane in

The abundance of this resource in China cannot be considered in isolation from the appalling safety record of the Chinese coal industry. Nevertheless, there is active interest and a number of foreign companies, including Phillips Petroleum, have provided advice on how China might utilise this reserve and possibly reduce oil imports.

China, role in the current international crude oil market

China, a hugely populous country, has major oil and gas reserves and an action plan for energy needs largely implemented by **Sinopec**. The country is a net importer of oil and is also undergoing a development program which can only increase the demand for imported oil. There is hardly any product which does not involve direct or indirect energy costs and the recent surge of development in this massive country caused a blip in oil supply and demand, which in the minds of some analysts was linked to the mid-2005 steep rise in the price of crude oil.

It is expected that sources from the **Caspian Sea** will help close the gap between production of crude oil and demand for it in China over the next few years. It is hoped that the coal-to-liquids facility planned by **Shenhua** will eventually provide about 1 million **bbl** per day of fuel from coal liquefaction, thereby reducing imports by an equivalent amount.

China Aviation Oil (CAO)
Importer of jet fuel for China, most of it from refineries and terminals in Singapore where CAO have a significant presence. The company generated large losses in 2004 by not anticipating rises in the price of crude oil. Since then BP have obtained a stake in CAO. The company South China Blue Sky Aviation Oil also imports jet fuel into China, and supplies fuel to fifteen airports.

Chinchilla, Queensland, Australia
Location of a trial **underground gasification of coal (UGC)** facility, from which gas is used to generate power at 67 MW. If it becomes commercial it will be the only source of electrical power of its type in the western world. Factors influencing its long-term viability include the price of natural gas: that was the experience of UGC developers in the US 20–30 years ago.

Chlorinated hydrocarbons
These burn in such a way that chlorine in the organic structure is converted to hydrogen chloride, for example in the combustion of ethyl chloride:

$$C_2H_5Cl + 3O_2 \rightarrow 2CO_2 + 2H_2O + HCl$$

Hydrogen chloride is of course very corrosive and its presence in the flue gases from the combustion of chlorinated organics is a problem. Moreover, small amounts of **dioxin**s will also be present. Flammability limits—upper and lower—for chlorinated hydrocarbons are similar to those for the parent, unsubstituted hydrocarbon. Liquid chlorinated organic waste can be disposed of by burning as droplets, similarly to fuel oils, in an incinerator.

Chlorofluorocarbons (CFCs)

CFCs are a family of inert compounds used for a variety of purposes. Chloroform is mainly used to produce chlorodifluoromethane (Fluorocarbon 22) by the reaction with hydrogen fluoride:

$$CHCl_3 + 2HF \rightarrow CHClF_2 + 2HCl$$

This compound is used as a refrigerant and as an aerosol propellant. It is also used to synthesise tetrafluoroethylene, which is polymerised to a heat resistant polymer (Teflon®):

$$2CHClF_2 \rightarrow CF_2=CF_2 + 2HCl$$

Carbon tetrachloride is used to produce chlorofluorocarbons by the reaction with hydrogen fluoride using an antimony pentachloride ($SbCl_5$) catalyst:

$$CCl_4 + HF \rightarrow CCl_3F + HCl$$

$$CCl_4 + 2HF \rightarrow CCl_2F_2 + 2HCl$$

The resulting mixture is composed of trichlorofluoromethane (Freon-11) and dichlorodifluoromethane (Freon-12). These compounds were used as aerosols and refrigerants. Due to the depleting effect of chlorofluorocarbons (CFCs) on the ozone layer, the production of these compounds has been reduced appreciably. Alternative compounds have been found with similar properties but with little or no effect on the ozone layer. Among these are HCFC-123 ($HCCl_2CF_3$) to replace Freon-11 and HCFC-22 ($CHClF_2$) to replace Freon-12 in such uses as air conditioning, refrigeration, aerosols and foam. These compounds have a much lower ozone depletion value compared to Freon-11, which was assigned a value of 1. Ozone depletion values for HCFC-123 and HCFC-22 relative to Freon-11 are 0.02 and 0.055, respectively.

Chocolate Bayou, TX

Scene of an olefins plant producing ≈ 1 Mt yr^{-1} per year for further processing. There was a fire in the summer of 2005, the sole casualty being a company fire fighter who suffered from smoke inhalation.

Chooz, France

Scene of the world's largest nuclear electricity generation facility. There are two reactors each capable of delivering 1455 MW of electricity. Chooz was in fact the site of the first ever nuclear power station in that region of the world, known as the Belgium-French Power Station, which came on line in 1967 and was decommissioned in the early 90s. Its output was 300 MW of electrical power.

CHP
See **combined heat and power**

Christmas tree
An assembly of valves, spools and fittings for an oil well to prevent the release of oil or gas into the environment and also to direct and control the flow of fluids from the well. When the well is ready to produce oil or gas, valves are opened to allow the fluids to flow through the **pipeline** to a refinery or storage vessel.

Originally named for its resemblance to the decorated tree of the same name, modern Christmas trees are so complex they no longer bear any resemblance and so are now often referred to as either a subsea or surface tree.

Chuchupa field
The only gas field in Colombian waters. It is operated by Chevron and supplies 75% of the natural gas used in Colombia. The heavy dependence on this one source arises partly from the fact that at the abundant oil fields of Colombia it is the practice to re-inject the **associated gas**.

Church of Latter Day Saints, role of in the Utah coal industry.
In the mid nineteenth century, the Church of Latter Day Saints made its base in what is now the State of Utah and became very powerful in the region. One of its policies was that the wood resources of Utah should be diverted to building rather than fuel use, which led to a demand for coal as an alternative to wood as fuel. The first discovery of coal in Utah was at Chalk Creek, later renamed Coalville, and mining began there. A better deposit was later found near Coalville at Grass Creek, and the mine there became known as Church Mine. Once Church Mine was 'up and running' coal production at Coalville ceased. Coalville was then built up into a residential town, which it still is. Coalville's development as a town also owed something to the Zion Cooperative Mercantile Institution, which was the brainchild of Brigham Young himself and is believed to have been the first chain of department stores in the US. Church Mine is still producing, now being part of the Southern Utah Fuel Company.

Citgo terminal, NJ
Scene of the leakage of 700 **bbl** of oil in 2004 when a tanker approaching the terminal struck a submerged object. An oil slick over 20 miles long in the Delaware River resulted.

Citrus peel

Orange and lemon peel have been used in places including Florida as a fuel. Citrus peel is capable of raising at least its own weight of steam. Citrus peel has also been blended with **municipal solid waste (MSW)** in **combustion** applications. The composition of MSW is variable and can to a limited extent be controlled by varying the amount of citrus peel with which it is blended.

City diesel

Diesel for vehicular use, having been treated with hydrogen to remove some of the sulphur as H_2S in order to reduce sulphur dioxide emissions. City diesel was developed in Sweden and introduced to the market in 1992. The treatment with hydrogen also removes polar and aromatic compounds to an extent which does not adversely affect the cetane number. Viscosity is however affected, and fuel pump malfunctions have been experienced in the use of this fuel.

Clair field

The most recently utilised (operations commenced April 2005) part of the North Sea oil reservoirs, with an eventual production target of 60 000 **bbl** per day. The field was discovered in 1978, but its exploitation had to await developments in **underbalanced drilling**. Clair field is expected to yield at least 5 billion bbl of oil. Its **block** numbers are 206/7, 206/8, 206/12 and 206/15. The US company Amerada Hess has a 9% holding in the field: they have a larger holding in the **South Arne field**.

Clarkston gas explosion

Occurring in a shopping precinct in Glasgow, Scotland in 1971, this explosion killed 20 persons and injured about 100 others. At that time the UK was changing from manufactured gas to natural gas. It was manufactured gas that had leaked, which is powerfully destructive on explosion due to its hydrogen content.

Clinker

Coal ash having descended to the base of the grate while molten, collecting unburnt or partially burnt coal particles during its descent, hence its black colour. It is used in road construction.

Coal, nature of

A combustible, carbonaceous family of sedimentary rocks formed from the compaction of partially decomposed plant material, a process known as **coalification**. The coal family includes **lignites**, sub-bituminous, **bituminous**, and **anthracite** and are classified according to their **coal rank**.

Coal, blending of

The blending of two coals of different origin is becoming increasingly common especially in the burning of coal as **pulverised fuel (PF)** in power generation. A **bituminous coal** can be blended with another such coal or with a **sub-bituminous coal**. Experience in several countries is that judicious blending of coals has major benefits in utilisation. For example, a high-sulphur coal can be blended with a low-sulphur coal to bring SO_x emissions to acceptable levels. Similarly, a coal with too high an ash content to be fired alone can be utilised if blended with a low-ash coal in such proportions as to bring the total ash yield to within the allowable range. A blend of two coals will be milled as such to particles in the PF size range, and energy required for the milling of a blend can be estimated from its **Hardgrove index** just as it is for a single coal.

A notable scene of coal blending is the **Nanticoke Power Plant** in Ontario.

Coal, price trends in

In the table below is the price variation of a ton of US **bituminous coal** over the period 1900 to 1970. The figures are averaged over the entire national production which was 212 Mt in 1900 and 602 Mt in 1970. The source of the information is the US Bureau of Mines.

Year	Cost in $US per ton
1900	1.04
1910	1.12
1920	3.75
1930	1.70
1940	1.91
1950	4.84
1960	4.69
1970	6.26

Coal auger

The very sharp rise between 1910 and 1920 did not occur until after 1916, in which year the price was $US 1.32 per ton. The jump in the 1940s occurred at about the middle of the decade. Clearly these sharp increases can be attributed to the two World Wars.

In moving 35 years on from the figures in the table we can use the current **CAPP** price as a comparison. It must be emphasised that this is for a high quality coal and not all of the numerous coals from which the averaged figures in the table were derived from would have conformed to the CAPP specifications. A typical CAPP price for 2007 is ≈$US 47.

Coal auger
Axially rotating screw device which, when applied to a coalface, cuts into it and conveys the coal pieces along the screw structure for direct transfer by gravity into a truck or storage container.

Coal bed methane
Methane initially adsorbed on to the surface of coal in a deposit. As the deposit is mined such methane is released and is a hazard known to miners as **firedamp**. Methane so released is however also a fuel resource. In the UK, there are many closed coalmines and these are known to be releasing methane which is, of course, a powerful greenhouse gas. There is some interest in obtaining such methane for use, but more promising is the production of methane not from mine voids but from coal still present in a disused mine. It is believed quantities of methane sufficient to be economically added to North Sea gas for distribution can be obtained by this means.

In the US, there is activity in coal-bed methane in some areas, including Virginia, with **CONSOL Energy** the main participant. There is activity in drilling first vertically then horizontally (a procedure previously used in the oil industry at sites including **Wytch Farm**) to access methane-rich coal. It is further intended that the space created by drilling for access to the coal could be used as a disposal site for carbon dioxide, making gas production and **carbon sequestration** an integrated process. In general, coal bed methane is an under-utilised resource awaiting investment. Even so, 8% of the methane needs of the US were met by coal bed methane in 2002.

Some care with terminology is required. The term coal*mine* methane is sometimes preferred for methane having already found its way out of the coal deposit as in a disused mine. The term coal *bed* methane is restricted to methane having been obtained by accessing the deposit to release it. If the term coal bed methane is used more broadly as in the earlier parts of this entry, methane obtained by accessing the coal and releasing the methane might be referred to as virgin coal bed methane.

Coal cleaning

Removal of non-combustible matter from coals high in such substances as mined. A common approach uses the fact that ash-forming constituents in the coal, being composed of mineral matter, are denser than the organic coal substance. Hence if the coal as mined is divided into smaller particles and placed in an aqueous medium there will be separation. The particles rich in coal will form floats and those richer in mineral matter will form sinks. There will also be middlings, which tend to remain suspended. It is possible to reduce the mineral matter content by a factor decided in advance by choosing the specific gravity of the aqueous medium which, of course, is adjustable by varying the nature and quantity of the solute.

Coal Clough wind farm

Located near Burnley Lancashire (previously a coal mining area), and producing power since 1992. Its capacity is just under 10 MW and it uses turbines manufactured by **Vestas**.

Coal dust

A frequently used though quite arbitrary definition is that coal particles which will pass through a Number 20 sieve, opening 0.51 mm, comprise coal dust while particles retained by such a sieve are loose coal.

Coal rail cars, corrosion of

Steel for the fabrication of rail cars for coal haulage is usually carbon steel. This has however led to corrosion difficulties especially where the coal being carried is wet and high in inorganics: such inorganics dissolved in water exit the coal particles and corrode the steel. A possible alternative is use of aluminium instead of mild steel. Over the last decade one particular US freighter of coal has been introducing rail cars made of S41003 stainless steel, which contains 1.5% nickel and up to 12.5% chromium. Results have been entirely positive, for example a life expectancy of 40 years for a rail car made from S41003 stainless steel in contrast to 10 to 15 years for one made from carbon steel.

Coal rank

The chemical classification of coal is a function of maturity. Coal rank increases from **peat** to **lignite**, sub-bituminous coal, **bituminous coal** and **anthracite**. Chemical changes occurring are aromatisation-condensation reactions of polymeric carbonaceous materials resulting in a gradual non-linear enrichment of **carbon** with increasing rank from lignite to anthracite, accompanied by a decrease in hydrogen, mineral matter and volatiles content.

Coal-biomass co-firing

Any **brown coal** will give a smaller output when used in power generation than a **bituminous coal**, due to its lower **calorific value** and (more importantly) higher moisture content, other parameters being equal. Therefore, any **biomass** which might be co-fired with coal will lead to performance reduction. It is for this reason that co-firing plants which had previously used only coal, such as that at **Fiddler's Ferry**, are restricted to biomass proportions of about 5%. Even so, in a power facility operating at GW level this represents very significant utilisation and enables the operator to comply with renewables obligations. With a biomass-only facility the target electrical power output can be linked to the fuel quality at the design stage and we can expect more such plants as **energy crops** proliferate.

When coal and biomass are co-fired, the intention is to obtain the same quantity of heat as with coal. An amount of biomass exceeding that of the replaced coal will be needed. The carbon dioxide emissions will be slightly *higher* than with coal only , but there will be a reduction in fossil fuel derived emissions. A further advantage to burning biomass is that biomass waste taken to a landfill leads in time to significant formation of methane, which is a much more powerful greenhouse gas than carbon dioxide. As energy crops proliferate the marginally higher total carbon dioxide emission from co-firing might be at least in part offset by their removal of carbon dioxide from the atmosphere. A further relevant point is that biomass waste if not co-fired or taken to a landfill might require simple incineration which will release CO_2 into the atmosphere with no benefit at all.

Coalbrook, Orange Free State, South Africa

Scene of a major coalmine explosion in 1960, in which 437 lives were lost.

This is two lives fewer than were lost at **Senghenydd,**

Glamorganshire in what was the UK's worst mining
accident of the 20th century. It occurred in 1913, at the
peak of UK coal production when over 230 000 were
employed in the industry.

Coalification

Process by which, on a time scale of tens or hundreds of millions of
years, vegetable matter is converted to coal. The first step along the
coalification sequence is microbial conversion of cellulose to humic
acids to form **peat**, which is the geological precursor to coal. At that
point, biological activity stops and the remainder of sequence is
geochemical, the first product beyond peat being **brown coal**, followed
by **sub-bituminous coal**, **bituminous coal** and **anthracite**. The degree of
advancement of a coal along the coalification sequence is the rank of
the coal, hence brown coal is a lower rank coal and bituminous coal a
higher rank coal. Alternative terminology would be that a bituminous
coal is more mature than brown coal. Within a classification such as
brown coal and bituminous coal, there are differences in maturity and
these can be quantified by reflectance of **vitrinite**, one of the group of
macerals of which coals are composed.

Factors which advance a deposit along the coalification sequence
are temperature and time and, to a lesser extent, possibly pressure.
Time alone will not suffice, and in the British Isles there are coals of
about the same age having significantly different rank, as determined
by **vitrinite reflectance**, because of the different geological conditions.
Thermochemical effects during coalification have led to temperatures
of up to about 300°C during the regular geochemical processes. In
some deposits there is evidence of much higher temperatures during
coalification because of incursion of volcanic material at some stage
in the sequence. In Scotland, there is a deposit in which the coal re-
sembles **coke**, having been carbonised *in situ* by such an effect.

Advancement in rank is accompanied by an increase in carbon
content. A typical brown coal might have 65% carbon and a typical
bituminous coal would have about 85% (figures in each case being for
dry coal). Also volatile matter, determinable by **proximate analysis**,
decreases along the coalification sequence. **Devolatilisation** of a brown
coal at temperatures in the region of 800–1000°C would involve about
70% weight loss whereas that of an anthracite at the same temperature would involve only a few per cent weight loss.

In addition to the geological maturity of a particular coal (expressible as its rank), the nature of the plant material from which the coal was formed is also a determining factor in its characteristics, expressible as coal type. A coal deposit is heterogeneous—banded—and the different bands are due to different sorts of initial plant debris. Microscopic examination of a section taken from a single band can reveal whether its origin was wood or bark and also possibly the plant species. The plants from that period were rather more basic in their functions than those which evolved from them which a present-day botanist would be familiar with. In particular, coal-forming plants propagated spores very wastefully, only a minuscule proportion of spores ever taking effect in reproduction. The remainder retained their structure over the coalification period so that to a trained coal petrographer they are quite recognisable under a microscope. Coals in which no plant structures are recognisable are called amorphous coals. Analysis of coal on the basis of different bands was the forerunner of the more rigorous procedure of **maceral** analysis and the type of a coal is formally expressed in terms of such an analysis.

Coconut waste

A staple fuel in the Philippines, with significant usage in Thailand and Indonesia also. It consists of shells and fibre and when directly burnt has a calorific value comparable to that of seasoned wood. It can be converted to **producer gas** by reaction with air and steam or to coco gas by partial gasification which yields as a by-product a **char** which can be made into **briquettes** for cooking applications. Such briquettes are in fact exported from the Philippines. Coco gas has a calorific value in the neighbourhood of 10 MJ m^{-3}, about twice that of a producer gas.

Cogdell Canyon Reef field

Onshore oil field in Texas, which after 50 years of usage was seen as no longer economic to work. Occidental, whose specialities include the application of **enhanced oil recovery** methods, acquired the field. Using carbon dioxide injection, Occidental were able to extract a further 20 million **bbl** of crude oil from the field.

Coffee waste, fuel use of

Fuel use of such wastes is not new but the engineering has tended to be primitive, for example spoilage of the coffee husk fuel by erratic drying practices. Currently in El Salvador there is very promising development work into power generation from coffee husk and pulp.

Independent tests on coffee waste as a fuel for power generation have been conducted by a UK combustion plant manufacturer, and generation at 0.5 MW level has been achieved straightforwardly.

Coke

The solid product of carbonisation of **coal**. The initial incentive for producing coke, quite early in the industrialisation of the world, was the need for a metallurgical reductant to produce iron from its ore. Previously charcoal had been used but trees became depleted over time and there was development work into making an alternative material from coal. All coals can of course be carbonised but not all yield, as a solid product of the carbonisation, a substance having the mechanical strength needed for metallurgical use in which the reductant has to support its own weight and that of the ore. **Lignites**, for example, do not yield a coke when carbonised but a **char**, which although chemically suitable for metallurgical use do not have the necessary hardness and cohesion.

Coking coals are usually chosen from bituminous coals, some of which are more suitable than others for coke production. There are laboratory tests—the Gray-King assay (see **caking**) and the crucible swelling number—by means of which suitability of a particular **bituminous coal** for coking can be assessed. The former gives an indication of the propensity of the coal to soften on heating, enabling particles to coalesce. The latter gives an indication of the extent to which a coal particle will expand on carbonisation.

The mechanism of coking is that volatiles released on heating act as a plasticiser and enable layers in the structure of the coal to slip across each other thereby making the coal fluid and coalescent. Later in the heating such volatiles are driven off and the fluidity is lost, a hard carbon product remaining.

In addition to its use in extracting iron from its ore, coke finds application as a fuel. Because it has been denuded of volatiles during manufacture, it burns with a red glow almost without a flame. Coke can also be gasified. Much of the **bituminous coal** which China produces is suitable for coking. At the present time, notwithstanding the difficulties with mine safety, China produces 56% of the world's coke.

Quality coke sells for about $US 155 per tonne: that is the price currently (1st quarter 2006) paid by Japanese importers of coke from Australia.

Coke breeze

Small pieces of carbonised coal produced in a coke oven and having slipped through the grate of the oven at its base. Coke breeze tends to be higher in ash-forming constituents than the larger pieces of **coke** which remain above the grate even though of course the two have both been produced from the same charge of coal. Coke breeze can be used as a fuel of a gasification feedstock.

Coke oven

Plant for the manufacture of **coke**. The coal to be coked is placed in an enclosure and supported on a grate. There are two types of coke oven: the by-product type and the beehive type. The latter are, in the 21st century, very few and far between. In a by-product coke oven, heat is applied and low-temperature volatiles displace air in the voids between the coal pieces providing an inert atmosphere whereby coking can take its course without erosion of the charge of coal by oxidation. If the coal has been ascertained, by prior laboratory testing, to be powerfully swelling the quantity of coal admitted to the coke oven has to be correspondingly reduced so that swelling will not cause breakage of the oven walls. Coke ovens range in peak operating temperatures from 750–1200°C. The higher the coking temperature the more complete the coal **devolatilisation** and therefore the less reactive will the final product be towards oxidation. Coking yields both liquid and gaseous by-products, respectively tars/oils and coke oven gas.

In a by-product coke oven provision is made for the tars/oils to be recovered for subsequent use. The coke oven gas, which has a **calorific value** just over half that of natural gas but a higher reactivity than natural gas, due to its high hydrogen content, might well be used on site. Coking often takes place at steelworks so that coke is produced where it is needed, there being no transportation costs. In such a situation the coke oven gas once cleaned can be used as a fuel in steel processing.

Beehive coke ovens work by partial combustion of the coal in the same way that charcoal for iron production was once made from partial combustion of wood.

Cold fusion

Investigators at the University of Utah and the University of Southampton in March 1989 announced that they had brought about fusion of deuterium nuclei in a simple apparatus at room temperature using only a pair of electrodes. Some mayhem followed when independent investigators claimed to have reproduced the work and then had to retract. The most important evidence against the claim was obtained by independent examination at Massachusetts Institute of Technology of spectra of electromagnetic radiation which had accompanied the alleged fusion. Before the end of 1989 a panel of the US Department of Energy had dismissed cold fusion.

Coleman fuel
Gasoline blended with some higher boiling material such as naphtha to raise its flash point, typically used in appliances such as camping stoves.

Colliery
The site where coal is mined, and processed both below and above ground. It includes all preparation equipment and workshops.

Colnbrook, UK
Scene of a proposed waste-to-energy plant which once commissioned will receive 400 000 tonnes of waste per year. The heat produced will be used in a **Rankine cycle** to generate electricity at about 30 MW, which will be sold to the grid.

Combined heat and power (CHP)
Sometimes called co-generation. A thermodynamic cycle which enables the fluid at the conclusion of the work-producing step to have **quality of heat** by being at a sufficiently high temperature to transfer its heat on to something else, possibly water used in domestic heating. CHP has been carried out not only with traditional fuels such as **bituminous coals** but also with **municipal solid waste** and (in Malaysia) with **biomass** fuels.

Combustion, conditions for
The process of burning a fuel with an oxidant to produce heat and/or work. Combustion is a controlled process; substantial amounts of experience underpinned by chemical and physical laws is involved. Combustion is governed by the '3 Ts':
• Turbulence: the interaction between two fluid streams required to achieve intermixing of the two;
• Temperature: the energy required for the initiation of a chemical reaction (i.e. oxidation); and
• Time: the period for the reaction to reach completion.
Mixing the fuel with an oxidant (usually air) and igniting the mixture by heating it to sufficient temperature results in a flame with a specific volume dictated by the time for the reaction to complete.

Combustion Institute

An American non-profit educational institution set up in 1954 to promote and disseminate international research in combustion science. Its main activity is the biennial International Symposium on Combustion.

Combustor

The combustion chamber where fuel and oxidant are mixed and ignited. In gas turbines and jet engines, the combustor refers to the complete system of combustion chamber, burners and injection system.

Compaña Española de Petróleos SA (CEPSA)

One of only three companies engaged in the refining of oil in Spain at the present time. The first refinery on Spanish territory was set up by CEPSA in 1930 in the Canary Islands and it is still in operation, receiving for processing 85 000 **bbl** of crude oil per day at the present time. Repsol, formed in 1987 as a result of mergers, and BP are the other two companies with refining operations in Spain.

Condensate

Valuable hydrocarbon liquid products that are present in gaseous form in the reservoir but condense on the surface. Gas temperature and pressure are controlled during gas production to maintain a gaseous state and prevent condensates forming until required.

Condensate, policy of OPEC towards

Huge amounts of natural gas – associated and non-associated – are of course produced by **OPEC** countries and such gas is often accompanied by condensate. Condensate is frequently diverted to the liquid stream after separation of oil and gas and can be used as a component of automotive fuel. However, OPEC do not for auditing purposes see condensate and crude oil as equivalent and a member country cannot aggregate crude oil and condensate in its production figures when reporting to OPEC.

Congo, oil production by

Significant oil production since the mid-1970s. Recent production has been enhanced by a **deep-sea oil and gas** field recently developed by Elf. The neighbouring Congo Democratic Republic has oil reserves in the Congo river estuary and further out to sea. **Chevron** and Fina have significant interests in oil recovery in the Congo Democratic Republic.

CONSOL Energy

US company dating from 1864, now supplier of coal for power generation to the extent that two-thirds of the conventionally generated electricity in the US is obtained using CONSOL fuel products. Its assets include the **Enslow Fork** mine. The company is also active in **coal bed methane** development.

Continental Oil and Transportation Company

Forerunner to Conoco, now ConocoPhillips, set up in Ogden Utah in 1875 by Isaac Blake. He began by bringing kerosene from a **refinery** in Colorado to Utah for sale as illuminating oil. The kerosene was brought to Ogden by rail and dispensed into small containers for distribution to the general stores of the town where it was sold to householders. Blake then extended his business to San Francisco, and in doing so oversaw the construction of the first pipeline in California.

Continental Oil and Transportation Company was acquired by Standard Oil in 1885. Standard Oil relinquished it by court order in 1913 and between then and the depression years the Company set up over 1000 filling stations in the western States; the familiar glass attachment to a filling pump which enables the purchaser to watch the fuel as it is being pumped was an innovation of the Company at that time. The Company merged with Oklahoma-based Marland Oil in 1928 to form the Continental Oil Company (Conoco), and from then on was involved in production and refining as well as in marketing and distributing.

Cool flames

Combustion phenomenon such that flames having a very pale blue colour and a temperature peak of the order of 100 K above ambient temperature, occur in an oscillatory fashion. The blue colour is known to be due to excited formaldehyde molecules. The first reported observations were in the mid 19th century. Over the last 40 years there have been many experimental studies of cool flames in the oxidation of hydrocarbons and oxygenated hydrocarbons such as acetaldehyde. 'Knock' in a spark-ignition engine is believed to involve cool flames. Quite possibly the intended ignition in response to the spark is preceded by a cool flame brought about by the compression prior to the spark.

Copahue, Argentina

Scene of **geothermal electricity** production at 670 kW. Argentina leads south America in geothermal resources, but over the last decade there has been a swing away from power generation to direct use of the hot water, for example in melting snow to make a holiday resort at the foot of the Andes accessible all year.

Core

Term used in oil exploration for a section of the geological formation removed for examination. Such removal is achieved by replacing the conventional **bit** in the drill with one in the form of a hollow cylinder with cutting edges.

Coriolis meter

Classical device for measuring the flow rate of a fluid, which impinges on a rotating disc on entering the meter. It has only recently been applied to **custody transfer of crude oil** as a replacement for pressure difference (PD) devices. Of course, whatever metering device is used in custody transfer, a reliable result will not be obtained without effective mixing of the oil.

Corn alcohol

Ethanol for commercial use (including **gasohol** production) made from corn. The first stage is to break down the starch in the corn to sugars, including dextrose and fructose, which can be fermented to produce alcohol. Henry Ford never surrendered his loyalty to rural America, where he had received his upbringing, and promoted corn alcohol usage in motorcars as a means of bringing business to struggling farmers in the 1920s. He described it as "the fuel of the future", a view shared by the head of research at General Motors at the time. In the early 21st century, activity in corn alcohol for motor fuels is very strong and includes the manufacture of **E85**.

Corn residues

These are burnt on a scale of tens of millions of tonnes per year in the USA and in Canada. In these countries, pelletised corn waste is available for burning in stoves in homes as an alternative to firewood.

Corrib gas field

Non-associated gas field in the north Atlantic close to the west coast of Ireland. It is a deep-sea resource, drilling having been to depths of 3300 m, and will have a life expectancy of 15–20 years once fully developed. Processing of the gas will be onshore.

Coryton, UK
East of London, the scene of a major refinery operated by BP, in operation since 1953. The present refining capacity is about 65 million **bbl** per year and the refinery has the space to store 4 million bbl of crude. The Thames estuary location provides for **supertanker** access.

Cotton residues
Significant use of cotton residues as fuel occurs in countries including Paraguay where about 3×10^5 tonne are burnt annually, a small proportion being co-fired with **coal** in electricity generation. The **calorific value** of cotton residue is 14–14.5 MJ kg^{-1}.

Cove Point, MD
Scene of a terminal and storage facility for **liquefied natural gas (LNG)**, owned and operated by BP, on the shores of Chesapeake Bay. The terminal entered service in 1978 but was closed in 1980 when gas prices in the US dropped. It was brought back into service on an intermittent basis in 1995 and fully recommissioned in 2003, having a storage capacity of 180 million m^3. Some of the LNG it receives is from Trinidad.

Cracking
Any refining process that uses heat and/or pressure in the presence or absence of catalysts to decompose and recombine hydrocarbon molecules to produce more valuable **petrochemical** products. See **catalytic cracking**.

Crop waste, fuel use of in India
Wastes from a miscellany of crops including rice, maize and mustard are being used in parts of rural India to make **briquettes** for fuel use in cooking. Such fuel has the intrinsic advantage of **carbon neutrality**. It is also very economic; it is being used in school kitchens which previously used **liquefied petroleum gas** at twice the fuel cost. It also benefits the farmers from whose crops the wastes come. They would previously have burnt off the crop waste but now receive about 500 rupee (£6) for as much waste as can be gleaned from an acre of land (about a tonne).

Crude oil, price trends in since 1869
All figures for the price of a **barrel** of oil quoted in this entry are related to 2004 US dollars, that is, they have been adjusted for erosion

of the value of a dollar over the 136-year period considered. Moreover the prices given are not linked to any one **benchmark crude**.

Although volatile, the present price of crude oil is around $60 per **bbl**, which is approximately what it was in 1869. There was a dramatic drop over the few years that followed, and at the end of the 19th century/beginning of the 20th Century, prices were around $10 per bbl. The price of a barrel remained within the $10 to $20 range until 1973–74 when restrictions imposed by **OPEC** caused prices to shoot up instantly to excess of $35. By the end of the 1970s a barrel cost $30, rising steeply during the following three years to over $60 due to the Iran/Iraq war. After that, the price fell sharply to around $20, but then rose again as a result of the Gulf War to nearly $30. During the closing years of the 21st century, the price was about $12, which fell again after 9/11.

By 2004 prices were around $40. The average world price of a barrel of crude oil between 1869 and 2004 was $19.41. Narrowing the timescale, the average world price between 1947 and 2004 was $22.86. Whether the price will come down to these sorts of values again, only time will tell.

Crude oil, total amount produced
The Oil Depletion Analysis Centre in London (www.odac-info.org) give a figure of 944 billion **bbl** of conventional crude oil produced over the entire time of the oil industry, that is, since about 1870. This converts to an average production over that period of just under 20 million bbl per day, a quarter of the 2005 production rate. The annual world production was in fact 20 million bbl per day in about 1955.

Crude oil, world production of
The most reliable mid-2005 estimate is 80 million **bbl** per day, worth about $US 4 billion. In 1985 it was 60 million bbl per day. In 2005, the crude oil refining capacity summed across the countries of the world is believed to be 90 million bbl per day.

Crude swaps
Term applied to exchange of oil between countries for convenience in transportation and distribution. As an example, at present crude oil from the Caspian Sea is taken to the Caspian port of Neka on the coast of northern Iran for local use while an equivalent amount of Iranian oil is exported from southern Iran through the Persian Gulf as Caspian oil. Transportation costs are therefore saved, although a

swap fee of about $2 per **bbl** applies. The scale of this operation is about 350 000 bbl per day. Iran successfully negotiated similar arrangements with Iraq and signed the contract in August 2006.

Cuba, liquid fuel situation in

For many years Cuba has imported refined gasoline from Venezuela, as well as crude oil, in a total quantity of about 55 000 **bbl** per day. It is not widely realised that part of the Gulf of Mexico comprises Cuban waters, and there has recently been discovery of an oil field containing about 100 million bbl of crude oil in this part of the Gulf. With two Canadian companies as operators it is expected that oil production at the field will begin in 2006. Indian and Spanish oil companies are also involved: the US embargo on trade with Cuba precludes the involvement of US oil companies.

Cull wood

Dead or rotten wood with no value to the timber industry. The wood is removed from trees to improve their health and can be burnt as a fuel.

Culm

Solid waste from the mining of **anthracite**. It contains sufficient carbon to be burnt effectively in a **fluidised bed**, as at **Northampton, PA** in power generation.

Cushion gas

See **base gas**

Custody transfer of crude oil

When crude oil is being transferred from a vessel to an onshore terminal, or possibly from a production offshore installation to a storage installation as at the **Fairly Baram platform**, a quantitative measurement of amounts of oil transferred is required as is knowledge of the amount of sediment, for pricing purposes. During such a measurement the oil is made to circulate a loop which causes the sediment to become suspended. In its simplest form, the loop is an inverted U-structure which the crude oil first ascends and then descends. Flow rate measurements before ascent and after descent provide a route to calculating the sediment content. There are a number of standards, including one by **ASTM**, for custody transfer measurement of crude oil.

Cutter field

Condensate field in the southern North Sea, capable of producing 3 million m^3 per day of gas which is taken to the facility at Bacton, Norfolk. The production platform uses wind power and **photovoltaic cells** which as well as avoiding carbon dioxide release will reduce the **EROEI** of the gas/condensate. The field, which has a life expectancy of about 15 years, is operated by Shell and ExxonMobil.

Czech Republic, coal production in

This country is a significant producer of coal with an annual output of about 60 Mt. **Bituminous**, **sub-bituminous coal** and **lignite** are all produced. Some of the **bituminous coal** in the Czech Republic is used to make **coke** for the country's iron and steel industry while a great deal is burnt as **pulverised fuel** to make electricity. About 15% of the country's total coal production is exported.

D

Dabhol project

Commissioning of plant for thermal generation of electricity in the Maharasthra region of India. The initial output will be 740 **MW**. Natural gas, purchased as **liquefied natural gas (LNG)**, is to be used as fuel. There have been uncertainties and delays in starting up the plant which, at the time of going to press, has been renamed the Ratnagiri project, due to a change in ownership. Meanwhile LNG is being sourced in Qatar, Oman and Australia. There is also the possibility that Egyptian LNG re-exported from **Bintulu** will be used.

Dahej terminal

Liquefied natural gas (LNG) terminal on the west coast of **India**, operated by **Petronet**. Its facilities include a regasification unit which enables gas to be reticulated to industrial zones. The Dahej project involved co-operation with the supplier in Qatar whereby they obtained two extra LNG tankers to ensure sustained supply to the terminal. Expansion of the liquefaction facilities at **Qatar** were also required.

Dakota Ethanol

Company in Wentworth, SD, producing 48 million US gallons of **corn alcohol** annually. Sixteen million bushels of corn are bought each year from local farms. The parent company owns ten other such plants in different parts of the US.

Dalton, GA

A major carpet manufacturing town in the US where carpet scraps, previously sent to a landfill, are used as fuel for one of the carpet factories. A quantity of 16 000 tonnes per annum of such waste used in this way will save $2.5 million per year in fuel oil costs. The combustion plant necessary to burn the waste fuel is expected to have paid for itself after four years.

Damietta

Egyptian port, close to the scene of a collision involving two oil tankers in 2005. A quantity of 9000 **bbl** of crude oil was released into the Mediterranean, and there had been a similar accident nearby only two months earlier.

Daura Refinery, Baghdad

Refinery having come into operation in 1955, one of the biggest in Iraq. Its capacity is 110 000 **bbl** per day although it is running at a third of that or less at the present time. Heavy fuel oil produced at Daura is currently being transported to Jordan by tanker for use in power generation.

Dead Sea, bitumen in

When salt is extracted from the waters of the Dead Sea, **bitumen** is also obtained. There is reference to this in the Old Testament, and in modern times there has been recovery and use of the bitumen in Jordan.

Dead storage

Coal that is stored in a compacted pile to prevent weathering. Requires reclaiming prior to utilisation.

Decentralised energy

Energy plant operating on a small, local scale, e.g, **wind turbines**.

Deep gas

Natural gas sources at depths greater than can be reached by conventional extraction processes, typically > 15 000 ft (4500 m). Deep gas is also the name given to a theory which states that primordial material buried deep within the earth is the origin of natural gas deposits. As the hydrocarbons migrate upwards they fragment into methane gas, which can then be extracted. Deep gas theory is an alternative to the origins of natural gas from living organisms.

Deep-sea oil and gas

As the offshore oil and gas industry has expanded, progressively deeper parts of the seabed have become accessible. In 1970, the depth limit for marine drilling was about 600 ft (180 m). Successive generations of **semi-submersibles** have provided advances on this and drilling to depths of 5000 ft (1500 m) is now possible. The result is the opening up of some reserves, the exploitation of which was previously precluded by their depth under the sea. Oil and gas from these is called deep-sea oil and gas. The cost of drilling at the maximum depth possible is about $US 0.4 million per day.

The first deep-sea oil from the North Sea was obtained from the **Alba field**. The first deep-sea oil and gas activity in Asian waters is currently taking place at **Kikeh,** Malaysia, where production is expected to begin in 2007. Risers from the sub-sea wells will take the oil to a **floating production, storage and offloading** as will be the case at the **Girrasol Field**.

Delabole

Cornish village, scene of the first commercial wind farm in the UK, operational since 1991 and now producing power at 4 MW. There are seven wind farms in Cornwall, with a combined capacity of 40 MW, equivalent to 4% of the UK total.

Demolition, waste wood from

Such wood waste is often taken to a landfill but there is increasing interest and activity in its use as domestic fuel after shredding. As such it serves as a benchmark against which the combustion performance of recently grown wood can be assessed. Waste wood from a demolished house has had decades (or in a few cases, centuries) in which to season. If recently gathered wood performs less satisfactorily than demolition waste in the same burning appliance, that is evidence that the former has not been seasoned for long enough before being used as a fuel.

In many advanced countries, including Australia and **New Zealand**, suitably treated timber is used for the walls of housing itself, instead of stone or brick. Timber housing is less common in the UK.

Demulsifiers

Chemical agents for the separation, for re-use, of waste oil from a two-phase mixture of oil and (usually contaminated) water. Garage waste is suitable for such treatment. There are also applications to offshore operations, for example separation of oil which has entered the seawater ballast in a **semi-submersible** so that when released back into the sea the ballast does not cause oil pollution. A mixture will typically require 2000 ppm of the demulsifier for effective use.

Denmark, oil reserves of

These are entirely in the Danish sector of the North Sea, where 9.4 billion **bbl** of crude oil are known to be in place. Like the UK and Norwegian sectors of the North Sea, the Danish started to produce oil in the early 70s. Most of that currently produced in the Danish sector is exported to other western European countries.

Denmark, wind power in

The entire electricity consumption of Denmark is only 3.7 GW, which could easily be met by wind power alone. Projects under plan are intended to provide wind power equivalent to 80% of the national requirement. **Vestas,** a leading manufacturer of **wind turbines**, are based in Denmark, so local knowledge is available. However, the intermittent nature of wind power capacity has to be offset by previously constructed inter-connections with Norway and Sweden.

Densification

The process used to compress **biomass** into **briquettes** or pellets to give a higher calorific value.

Derrick

The complex support for drilling tools, casing and pipe and apparatus for raising and lowering equipment on an oil rig or over a borehole. Named after Thomas Derrick, an Elizabethan hangman who used a similar support structure from which criminals were hanged.

Desulphurisation

Flue gas desulphurisation (FGD) technologies are used to control the emissions of SO_2 and SO_3 from coal- and oil-fired power stations, refineries and metallurgical plant. The widespread adoption of FGD will significantly reduce anthropogenic sulphur emissions worldwide and help improve air quality to benefit human health and the environment.

Sulphur is one of the most common elements in the Earth's crust and occurs widely as an impurity in coal, crude oil and many ores. Global SO_2 emissions amount to approximately 140 Mt yr^{-1} (~ 2 Mt yr^{-1} in the UK). SO_2 emissions can be controlled in several ways:

- Switch to a fuel or ore with lower S content
- Improve efficiency of the process to use less fuel
- Remove the S before use (usually impractical)
- Remove S during use
- Remove SO_2 from flue gases before release to atmosphere.

The final option is the most efficient means, and FGD technologies have been developed to this end.

SO_2 is acidic in nature so removal from flue gases can be achieved by reaction with a suitable alkaline material, such as **limestone** (essentially $CaCO_3$), quicklime (CaO), or hydrated lime ($Ca(OH)_2$). The reaction with SO_2 produces a mixture of sulphate and sulphite salts of calcium (in the above cases). In some processes all sulphite salts are converted to sulphate. In wet FGD systems, a solution or slurry of alkali meets the flue gas in a spray tower. The SO_2 in the gas dissolves in the water to form a dilute acid which then reacts with, and is neutralised by, the dissolved alkali. The salts then precipitate out of solution. In dry and semi-dry FGD systems, a solid alkali sorbent is either injected or sprayed into the flue gas stream, or the flue gas passes through a bed of alkali. The alkali must be porous or finely divided for this to be effective. In semi-dry systems, water is added to the flue gas to form a liquid film on the particles in which the SO_2 dissolves, promoting the reaction with the solid. There are a wide range of FGD processes available. Six processes commonly operated around the world are described below.

1. *Limestone gypsum* A wet scrubbing process in which the flue gas is treated with a limestone slurry in order to remove and neutralise SO_2. The final product is gypsum ($CaSO_4.2H_2O$) which can be sold to make plasterboard. This process is favoured in the UK at large power stations such as **Drax** and Ratcliffe due to the limestone deposits to the south of Buxton in Derbyshire. The reactions are:

Absorption $\qquad\qquad SO_2 + H_2O \rightarrow H_2SO_3$

Neutralisation $\qquad H_2SO_3 + CaCO_3 \rightarrow CaSO_3 + CO_2 + H_2O$

Oxidation $\qquad CaSO_3 + \frac{1}{2} O_2 + 2H_2O \rightarrow CaSO_4.2H_2O$

2. *Sea-water washing* This process uses the natural alkalinity of untreated seawater to scrub the flue gas and neutralise SO_2. The water used is treated with air to reduce its chemical oxygen demand and acidity and then discharged back to the sea.

3. *Ammonia scrubbing* This process is similar to the limestone gypsum process except that aqueous ammonia is used as the scrubbing agent.

The SO_2 is removed from the flue gas by reaction with ammonia to give the product ammonium sulphate, a valuable fertilizer.

$$2NH_4OH + SO_2 \rightarrow (NH_4)_2SO_3 + H_2O$$

$$2(NH_4)_2SO_3 + O_2 \rightarrow 2(NH_4)_2SO_4$$

4. *The Wellman-Lord process* This is a regenerative process which involves the wet scrubbing of SO_2 with sodium sulphite solution. A saleable by-product is produced; either elemental sulphur, sulphuric acid or liquid SO_2 depending on plant design.

$$SO_2 + Na_2SO_3 + H_2O \rightarrow 2NaHSO_3$$

$$2NaHSO_3 \rightarrow Na_2SO_3 + SO_2 + H_2O$$

For sulphuric acid as the final product, the reactions involved are:

$$SO_2 + \frac{1}{2} O_2 \rightarrow SO_3$$

$$SO_3 + H_2O \rightarrow H_2SO_4$$

5. *Circulating fluidised bed* In the CFB, the flue gas is passed through a dense mixture of lime ($Ca(OH_2)$), reaction products, and sometimes **fly ash**, which removes the SO_2, SO_3 and HCl. The final product is a dry powdered mixture of calcium compounds.

6. *Furnace sorbent injection* This process involves the injection of hydrated lime into the furnace cavity of the boiler to absorb SO_2. Spent absorbent is extracted with the fly ash in the **electrostatic precipitator**. This process is cheap to install and run, and can remove up to 70% SO_2 from the flue gas.

Development well
An oil well drilled in the expectation that it will become a production well, having previously been the subject of a geological survey and, possibly, a prior appraisal well. A development well will also follow a **wildcat** which has given positive results. In the drilling of a development well the depth at which oil occurs will be known.

Devolatilisation
The decomposition of coal or other solid fuel by heat, yielding gases and vapours collectively called volatiles. Flammable volatiles are tars/oils, light hydrocarbons such as methane, **ethylene** and **ethane**, and **carbon monoxide**. Non-flammable volatiles, much less abundantly yielded than the flammable ones, are carbon dioxide and water. When a high-volatile fuel is burnt, devolatilisation occurs concurrently with combustion. Heat from the burning of the flammable volatiles feeds

back to the remaining solid to promote further devolatilisation. Devolatilisation of coals as a function of temperature can be represented by suitable **Arrhenius parameters**. In solid fuel applications, **pyrolysis** can be taken to be synonymous with devolatilisation.

Devolatilisation, being thermal degradation and not combustion, can occur in an inert atmosphere and many laboratory studies of devolatilisation of coal under inert conditions, for example in a reactor filled with argon, have been reported. Devolatilisation involves many reactions within the coal structure including methyl group removal, decarboxylation and cleavage of parts of the coal organic structure. The overall heat effect is the result of the individual heat effects of all such reactions occurring and might be net positive (endothermic) or net negative (exothermic). It will however, regardless of its sign in the thermodynamic sense, be small in comparison with the heat of subsequent *combustion* of volatiles so released.

Dew point

The temperature at which liquid will start to appear when moist gas is cooled. Its most common application is to moist air in air conditioning but it also relates to post-combustion gas, hence its inclusion in this volume.

Imagine post-combustion gas having cooled to 32°C. Its humidity is 52%. From tables, the pressure of water vapour at 32°C is 0.04754 bar. If the humidity is 52% the vapour pressure is then:

$$0.52 \times 0.04754 \text{ bar} = 0.0248 \text{ bar}$$

The temperature at which this is the saturated vapour pressure of water is the dew point. From tables this is 21°C, which is therefore the dew point.

Diesel, Rudolph (1858–1913)

Inventor of the engine associated with his name and which can thermodynamically be analysed by the diesel cycle. It differs fundamentally from the primary competitor of Diesel's own day in that the working substance is air, not steam. It also differs from the petrol engine in that ignition is due to heat generated by gas compression and not, as in a petrol engine, from a spark. By 1893 when Diesel obtained a patent for his device, oil refining was established. A boiling fraction much further along the boiling range of crude oil than gasoline was found to be suitable for this type of engine, and given the name diesel.

Throughout the 20th century, diesel engines for both stationary and mobile applications proliferated, the latter being chiefly heavy trucks and also of course trains. There was a swing to diesel for private passenger cars in the 1970s following the period when crude oil was withheld from the US by the **OPEC** countries. By 1980 many major manufacturers including Nissan and General Motors were offering diesel versions of their respective models. Towards the end of the 20th century, there was a decline in the use of diesel engines for passenger cars, and at the time of writing only Mercedes Benz and Volkswagen export diesel cars to the US. Even so, in the countries of the EU there is currently a shortage of diesel, with one projected solution being the **Porvoo refinery diesel project**.

Rudolph Diesel disappeared over the side of a ship on his way to England in 1913. It is not known whether it was accident, suicide or if he had been deliberately pushed. In support of the third possibility is the fact that his invention, by then well developed, was potentially very useful to the German Navy in the approaching Great War, and Diesel himself was very concerned about this. It might well have been that he was travelling to England with this issue as his agenda when his life ended.

Digboi

The scene of oil activity for over 100 years in the Assam region of India. A well was first sunk there in 1866 although it was not until 1889 that oil production began on a commercial basis, the beginning of the Indian oil industry. Such oil was taken to Margherita for refining. Refining at Digboi began in 1901 and has continued to the present time. It is often claimed that the refinery at Digboi is the oldest in the world.

Dioxin

A substance in the polychlorinated dibenzo-p-dioxin family of organic compounds. The plural—dioxins—is taken to include all members of the family of compounds. They are among the most harmful substances known and the release of a few kg constitutes a major incident. They form during the incineration of chlorinated wastes including PVC.

Directional drilling

A method of drilling for oil and gas off the vertical axis used to reach many parts of a reservoir from a single drilling point, or to reach an area where vertical drilling cannot be carried out such as under a shipping lane. Directional drilling is also used to intersect reservoirs at angles that expose more of the rock to the well bore and so increase the amount of oil or gas that flows into the well.

Disposal well
An oil or gas well, usually disused, into which waste fluids can be injected. For example: **carbon sequestration**.

Distillate-oil-to-gas
Term for gas made from petroleum distillate by **steam reforming**, used chiefly in the context of pricing of the various fuel gases. If the gasification feedstock was residual material from refining, the corresponding term residual-oil-to-gas applies. Such classifications will be compared on a $US per million **British Thermal Unit** basis with natural gas supplied either as such or as **liquefied natural gas (LNG)**. The **Henry Hub** price of natural gas might be used for guidance in such comparisons but will not apply strictly to anything other than pipelined natural gas. Pipelined natural gas and LNG are competitors, especially in the US domestic market: the Henry Hub price is a formal benchmark only for the former.

Distributed energy
Generation of electricity and/or heat at, or close to, the point of demand. This includes small scale **fuel cell** technology, **photovoltaic cells** and **wind turbines** to larger **combined heat and power** plants.

Djibouti
Country in northern Africa with a population of 634 000, located across the Red Sea from Yemen, bordered by Eritrea, Ethiopia and Somalia. It is an important regional centre for crude oil supply, being a Free Trade Zone and having at its port the space to store 1.26 million **bbl** of oil. At the port at Djibouti, **bunker fuel** for tankers is sold by Exxon, Shell and Total. Each of these also distributes petroleum products within the country, as does Chevron. Djibouti does not itself at the present time produce oil, the oil companies of the world having no interest in exploring Djibouti waters in spite of encouragement from the government. The country generates electricity at 85 MW, all of it thermally using fuel oil. It is however intended that there will soon be **geothermal electricity** in Djibouti: the viability of a 30 MW geothermal facility has been confirmed and finance negotiated to realise it.

Doha, Qatar
Site of a **gas-to-liquid fuels** plant owned jointly by Chevron and **Sasol** with an output of 100 000 **bbl** per day. It uses as feedstock non-associated gas from the **Qatar North Field**.

Dokdo

Two small islands about 215 km from the coast of South Korea, sovereignty of which has been claimed both by Japan and South Korea. There is currently exploration for **natural gas hydrates** at two sites in waters off Dokdo. One is being undertaken by the state owned Korea Gas and the other jointly by Korea National Oil and Australia's Woodside Petroleum. These are Korean endeavours and investments, and South Korea's Coast Guard ensures that no Japanese boat comes within 20 km.

Dominion Iron and Steel Company (DISCO)

Set up in 1901 in Sydney Nova Scotia from the amalgamation of previous coal and mineral organisations, soon becoming the largest steel works in the British Empire. The coke ovens at the steel works were supplied with coal from the Sydney Coal Field and iron ore was brought from Newfoundland. As a result of a merger in 1930, DISCO became part of the British Empire Steel Corporation (BESCO), which in 1957 was bought out by Hawker-Siddeley. By the 1960s it was believed that there was insufficient accessible coal in the Sydney field to sustain the local steel industry beyond the early 1980s. In the mid-1980s the coke ovens at the Sydney steel works were mothballed in favour of electric arc heating with graphite electrodes.

Double-hulled tankers

In oil tankers from the **Gluckauf** to some of those built recently, the interior of the hull is compartmentalised into a number of separate but adjoining tanks which have the ship's hull itself as an outside wall. In the latest generation of tankers this is not so: there is a second hull enclosing the first so that in the event of a collision this protects the inner hull, upon which containment of the oil depends. It has been decreed that from 2015 all oil tankers in US waters must be double-hulled, a decision partly brought about by the **Exxon Valdez** spill. Given the huge amounts of hydrocarbon liquids entering or leaving the US by sea each year, the frequency of collisions can never be zero. Once double-hulled tankers become the norm, however, the *probability* that a vessel having collided will leak its contents will be greatly reduced.

The EU has responded similarly; only double-hulled vessels will be permitted to enter its waters from 2010. A double-hulled **supertanker** collided with a freighter in **Groensund Strait** in 2001. The supertanker released 8% of its payload. Similarly in 2006 the double-hulled **Seabulk Pride** released only a few gallons of hydrocarbon during collision with a sheet of ice at an Alaskan coastal location in 2006.

Dounreay, Scotland

Although the power production at **Arco, Idaho** a few years earlier had in fact used 'fast neutrons', the plant at Dounreay in Scotland is usually credited with being the first to produce significant power from a **fast reactor**, that is, one in which neutrons impinge at high kinetic energy on U^{238} atoms and produce the fissionable Pu^{239}. It began power production in 1959 and generated electricty at 15 MW. It has, for a number of years, been under decommissioning. The site it occupies will not be declared totally safe until about 2036.

Dragon Bay field

In contrast to the **Morecambe Bay gas field** and the **Liverpool Bay oil and gas fields** the nearby Dragon Bay field has disappointed explorers for gas. The field was first explored in 1994 and activity by Marathon Oil continued until early 2006. In the event that the exploration had yielded positive results, it was proposed to bring the gas ashore at the Welsh coast where there are two **liquefied natural gas** facilities.

Drake, Edwin L (1819–80)

Individual credited with sinking the first oil well in the US in Titusville Pennsylvania in 1859. The depth of the well was about 70 feet.

Drax

The largest power station in Western Europe, Drax is located between Selby and Goole on the River Ouse in Yorkshire. Named after the local village, the name Drax comes from the old English meaning: 'A timbered patch of dry land haunted by beasts amid the marshes'. Drax produces enough electricity (4 GW) to meet the needs of four million people or six cities the size of Leeds. Completed in 1986, Drax can burn up to 36 000 tonnes of **coal** per day to produce electricity in six 660 MW generating units. The flue gas **desulphurisation** plant is the largest of its kind in the world.

Drift coal

Coal having been displaced from the site of its formation and therefore of the vegetation from which it was formed by **coalification**.

Drilling fluid

Used in the rotary drilling of oil wells, a function being to lubricate the **bit** and keep it cool. It also causes solids removed in the drilling to rise to the surface and enters vulnerable areas of the newly exposed wall, temporarily consolidating them until a metal casing has been installed.

Drilling fluids are aqueous in nature, their properties such as **viscosity** and pH being controlled by the precise composition. The simplest drilling fluids consist of clay and water in various proportions. Some have additional ingredients including barium sulphate which, containing a heavy element (atomic weight of barium = 56), will raise the density of the fluid and provide for protection in the event of unexpected hydrocarbon or water release in drilling. This is especially important in the drilling of a **wildcat** well where such contingencies are more probable.

Dry gas

Non-associated natural gas which also lacks **natural gas liquids**. Dry gas fields occur in the southern North Sea.

Dual-fired

Name given to a generating system that produces electricity from two or more fuels. Not all systems can use the two fuels simultaneously; these use the primary fuel for continuous generation with the alternative fuel being used for start-up.

Dung Quat

Scene of a proposed refinery in Vietnam. At present, in spite of producing a great deal of oil at fields including **White Tiger**, Vietnam imports refined products as there is no domestic refining capability. **Petrolimex** are the major importer of refined hydrocarbons into Vietnam. Plans for a **grass roots refinery** at Dung Quat began in 1995. There were setbacks and several overseas oil companies who had become involved in the construction of the refinery withdrew. There was also lack of confidence by possible investors, not least because Dung Ho is 1000 km from the White Tiger field from which most of its crude supply would come. The project has not been abandoned and commencement of production within the next 2–3 years is hoped for.

It is interesting to compare Vietnam with Japan. The former has huge amounts of crude oil and no refining capacity. The latter has enormous refining capacity and very little domestic oil. Yet Japan has long been a key industrial nation while Vietnam, which had a protracted war that ended 30 years ago, is still a poor nation. To draw facile conclusions from these isolated facts would be wrong, but it is perhaps legitimate to infer that oil resources without good management and planning are not sufficient to sustain a vibrant economy and high standard of living.

Duvha power station

Situated in Mpumalanga Province of South Africa, this power station commenced operation in 1975. Steam for the turbines is raised by means of locally won **bituminous coal**, and the capacity of the facility is 3500 MW of electricity.

E

E85

Automotive fuel comprising a blend of 85% **ethanol**, balance gasoline. It is routinely available in certain US states including Illinois. E10, having only 10% ethanol, is available in countries including the US, Canada and Australia. A vehicle not designed as a **Flexible Fuel Vehicle (FFV)** will often perform satisfactorily on E10.

Eakring field

The first oilfield in England to come into production, in rural Northamptonshire.

East China Sea

The scene of current exploration for natural gas. China and Japan dispute the exact position of the maritime borders.

Easy oil

Term which, although coined much earlier, has been used a great deal in discussing the US Energy Bill which has recently been signed by President Bush in New Mexico. It refers to oil straightforward to access, and such oil is believed to be declining in availability in the US. This means either importing more oil or accessing difficult oil domestically. An example of the latter is drilling taking place on the shallow-water shelf in the Gulf of Mexico. Reaching the seabed to drill is easy but accessing the oil requires a sub-sea well of depth 32 000 ft (9750 m). Creation of the well will take a year. Oil and gas in shallow waters requiring deep drilling for access has to be distinguished from **deep-sea oil and gas** where deep applies to the water depth, not the drilling depth.

Eccleshall Biomass, UK
Company currently developing an electrical power facility to use **elephant grass** as fuel. It will produce power at up to 2.2 MW and will provide business for the local rural area where the **energy crop** can be grown.

Economiser
A device used in a coal-fired power station to preheat feed water from the turbine condenser using flue gas exiting the system. Present day economisers are closely based on the original patented by Edward Green in 1845.

Ecuador, energy scene in
There has been oil production in this country since 1917 and at present just under 20 Mt yr^{-1} of crude oil and **natural gas liquids** are produced. Two thirds of it is exported. There was an extraordinarily successful exploration in 1993 as a result of which the country's proven reserves increased all at once by 25%. By contrast, natural gas is fairly sparse. Sadly, much **associated natural gas** is at present simply flared off although there are proposals to stop this practice and to make the gas available for domestic users. There is **non-associated gas** at a field in Ecuador waters which produces 0.8 million m^3 of gas per day, all of which is used to generate electricity which is actually consumed in Peru. Over 60% of Ecuador's own electricity comes from the hydro-electric facility at **Paute**, the remainder from thermal plants using oil. There are not insignificant reserves of **lignite** in Ecuador but these are not utilised at the present time.

Edmonton solid waste plant
The recipient of **municipal solid waste** and other sorts of waste (including trade waste) from several London boroughs. It is burnt to make generate electricity which, after the plant's own power requirements have been met, is contributed to the grid. Scrap tyres received at Edmonton are passed along to **Elm Energy and Recycling (UK) Ltd**.

Een gas field

Onshore natural gas field in the Netherlands containing only 15 million m³ of gas, little more than a millionth the amount in the **Groningen** field in the same country. A field such as Een would not have been economic to develop until the recent introduction of a device known as a skid, a gas wellhead plant which can routinely be moved from one well to another. It is expected that the Een field will be depleted after three months of production at which point the skid will be taken to another very small gas field in the Netherlands, of which there are several at places including Lauwerzijl in the north and Spijkenisse in the west. At Een the manufacturers and suppliers of the skid also drilled the well.

Efficiency

The effectiveness of a system based on output relative to input. For example, energy efficiency is measured as power output divided by power input, the result being a dimensionless number. When expressed as a percentage, the energy efficiency of a closed system cannot exceed 100%.

Egmond aan Zee

Scene of an offshore wind farm to be built 10 km from the Dutch coast. The wind farm will be 50% owned by Shell, the turbines having been constructed by **Vestas**. There will be 36 turbines with a combined power generating capability of 108 MW. Production of electricity began in 2007.

Egypt, hydrocarbon reserves in

It is believed that the basket of rushes in which the infant Moses was placed in the Nile had been smeared with crude oil to make it waterproof, according to the Book of Exodus. It is in any case certain that that long ago crude oil was used in that region of the world as a binding or waterproofing agent in building. In the opening years of the 21st century, Egypt has over 400 Mt ($\approx 3 \times 10^9$ **bbl**) of proven reserves of crude oil, about two thirds of which is offshore in the Gulf of Suez. Some of the onshore fields are in Sinai.

Development is ongoing and as recently as April 2005, an exploration well some 300 km southeast of Cairo drilled by Premier Oil gave very promising results.

Eilat-Ashkelon Pipeline

Oil pipeline linking the Red Sea port of Eilat with the Mediterranean port of Ashkelon. Entering service in 1968 its initial purpose was to convey Egyptian crude across Israel to the Mediterranean, some of it being diverted to Israel's two refineries along the way. A second parallel pipeline has, in one of the most strategic innovations in oil transportation that recent years have seen, come into use in the opposite direction. At Ashkelon this receives crude oil originating from the Former Soviet Union and from the **Caspian Sea** which has been shipped or piped to the eastern Mediterranean, including some from the **Baku-Tbilisi-Ceyhan Pipeline**, and transfers it across Israel to Eilat. From there it can be loaded on to tankers bound for Asia via the Indian Ocean. The ports of Ashkelon and Eilat can take vessels the size of a **supertanker**, which the Suez Canal cannot, and this is a significant advantage.

Ekofisk field
The first oil field in the Norwegian sector of the North Sea to enter production, in 1971. Note that the British and Norwegian sectors started to produce oil at around the same time.

Elbistan
Scene of the largest of the many **lignite** deposits in Turkey, the total recoverable lignite being of the order of 10^{10} tonnes. There is much utilisation of Turkish lignite to make electricity. Lignite from the Elbistan deposit has a **calorific value** in the bed-moist state of about 6 MJ kg^{-1}. Lignite combustion is increasing in Turkey. Lignites are more susceptible to self-heating in storage than higher rank coals and there is at the present time considerable research and development into this in Turkish coal laboratories.

Electricity prices, increases in for carbon sequestration
At the present time in the UK, electricity costs are around £40 per MWh. For thermally generated electricity, including removal of the carbon dioxide from the post-combustion gas followed by its burial under the sea, such as is occurring at the **Sleipner field**, costs would be between £7.50 and £15 per MWh. To some degree this cost is indirect. Some of the power produced would have to be diverted to the sequestration process, making less available for sale, resulting in a higher charge.

Electrostatic precipitator
Device used to remove particulate matter from an air stream by elec-

trically charging the particles and collecting them on an oppositely charged plate. Used in coal-fired power stations to collect **fly ash** from the flue gases before they are vented to the atmosphere.

Elephant grass
Synonym for **Miscanthus**.

Elizabeth Watts
The name of the first vessel to transport oil across the Atlantic in 1861. The vessel was not a tanker; it carried the oil in barrels.

Elk Hills oil field
Situated north of Los Angeles, this oil field has been producing for over 90 years. Initially oil from this field was allocated exclusively to the US Navy. It is now operated by Occidental who have applied **enhanced oil recovery** methods as they have at other places including the **Cogdell Canyon Reef field**. There are high levels of **associated gas**. The location is such that the field straddles the southern California and northern California oil and gas markets.

Elm Energy and Recycling (UK) Ltd
Currently the only plant in Europe which uses waste tyres as a fuel for the generation of electricity. It has the capacity to burn 100 000 tonnes of tyres per year. An estimate of the power produced is calculated below:

---

100 000 tonnes = 10^8 kg
Taking the **calorific value** to be about 35 MJ kg^{-1}
Heat released in a year = 3.5×10^{15} J
$\Downarrow$
1.1×10^8 W of heat
Taking the generation to be about 30% efficient, electricity is produced at 33 MW

The above approximate calculation has been checked against information on the company's web pages, which give the lower figure of 25 MW. In either case, significant power for sale to the grid is possible. Tyre suppliers, remoulders and distributors have business arrangements with Elm Energy and Recycling whereby scrap tyres are sold to them. Tyres tend to burn very smokily and a large proportion of **excess air** is necessary to reduce particulate to acceptable levels.

There is also limited activity in the use of **tyre derived fuels** for power generation in the US, though as a supplementary fuel rather than a primary one as at Elm Energy and Recycling (UK) Ltd.

El Salvador, energy scene in

This country produces no oil, natural gas or coal. It imports only oil, both crude at about 20 000 **bbl** per day and refined products including gasoline. There is major hydroelectric activity and also **geothermal electricity**: these account for about half of the country's electricity consumption, the balance being thermal. Coffee waste is also being used as a fuel on an exploratory basis. There are also significant quantities of **bagasse**. Largely through the efforts of international aid agencies, there is limited activity in renewable sources including the **wind turbine** and the **photovoltaic cell**.

The pipeline from Mexico to **Escuinla, Guatemala** might well be extended to the border with El Salvador, bringing natural gas to the country for the first time.

Ely Power Station

The scene of electricity generation at 38 MW using cereal straw as fuel. The fuel is used to raise steam which is then superheated to increase the efficiency of the Rankine cycle which it undergoes.

Emily Hawes field

Onshore gas field on a small island off the Texas coast in which the Houston-based Petrogen have a major holding. The field was discovered in 1984 although production did not begin until 2005 and is, at the time of writing, about 25 million m^3 per day. Petrogen had previously been explorers rather than producers of natural gas and see the Emily Hawes activity as their entry into production.

Emissions standards

Standards imposed by law on the release of pollutants such as SO_x, NO_x and particulate by organisations such as power utilities which burn fuels on a large scale. The emission standard depends on the nature of the fuel used. For example, it would be quite illogical to impose the same particulate emission standard on a plant using **bituminous coal** and on a plant using natural gas since the former has a higher intrinsic propensity to particulate formation than the latter.

An emission standard will often be on the basis of heat produced rather than weight of fuel burnt, for example the following emission standard taken from a US source:

1.2 lb SO_2 per 10^6 **British Thermal Units (BTU)** *of heat* for
coal combustion.

Heat is italicised to emphasise the point that in this standard, up
to 1.2 lb of **sulphur dioxide** can be produced per 10^6 BTU of heat released by the fuel, irrespective of how effectively that heat is later converted to work. In a typical **Rankine cycle**, 10^6 BTU of heat would
provide about 3.5×10^5 BTU of work at the turbine, but that is irrel-
evant to the standard as expressed. Gaseous pollutants are diluted by
a factor of about 10^3 on dispersion, at which stage **ambient standard**s
apply. The emission standard will not in general be the emission immediately after combustion but that after cleansing of the post-combustion gas, for example by removing some of the sulphur dioxide by
washing with water. The gap between the emitted amounts at the combustion plant and the emission standard will determine what extent
of removal of a particular pollutant is necessary.

Emissions trading

At the present time, most such trading is related to **greenhouse gases**, in
particular carbon dioxide. Organisations which burn fuels to produce
energy are assigned allowances on emissions of carbon dioxide in units
of tonnes. A company can emit in excess of its quota by purchasing
units from another which does not require its full quota. 'Carbon dioxide credits' have a sale value of between $US 3 and $US 5 per tonne of
carbon dioxide. A release of 4500 tonnes per week of carbon dioxide
above the quota would therefore cost about $US 1 million per annum
to offset by the purchase of credits. An advantage of emissions trading
is that purchase of units by an enterprise which has a carbon dioxide
release rate in excess of its quota is cheaper than alternative measures
such as using a different fuel or **carbon dioxide sequestration**.

In the UK, no CO_2-producing utility is obliged to participate in
emissions trading: it can operate perfectly legally by not exceeding its
own quota. By the same token any such utility has a statutory right to
buy or sell units on the emissions market.

Emulsion fuels

Term currently applied to fuels in which **diesel** is present as an emulsion in water, there also being a stabilising agent. Such fuels contain
10–20% water and their advantage is a reduction in emissions, in particular of oxides of nitrogen. Public service vehicles in places including Nottingham UK, and several Italian cities including Milan are
using such fuels. **Orimulsion®** is also such a fuel.

Energiewende
Term applied to Germany's current policy on energy, broad meaning 'transformation of energy systems'. It signifies Germany's plans to have wound down nuclear power generation by 2020 and to have replaced it with power from renewables, especially wind power in which the country is investing very heavily.

Energy
Formally defined as 'the ability to do work'. Work, in this sense, is the action of a force acting on an object undergoing a displacement. Energy can be transformed from one form to another. For example: a heat engine transforms chemical energy to thermal energy to mechanical energy (and then electrical energy if required by the system). A **battery** transforms chemical energy to electrical energy (and then to mechanical energy if required). A **fuel cell** transforms chemical energy to electrical energy (and then to mechanical energy if required).

Energy Africa
Based in Cape Town, this company, which was recently acquired by **Tullow Oil**, has a major stake in the **Kudu gas field** in Namibia. It also produces 18 000 **bbl** of oil per day from fields in Gabon. Energy Africa accounts for about half the assets of Tullow Oil.

Energy audit
An assessment of the energy usage of a home or building which reports on current usage of electricity, gas and water as well as the state of the building, making recommendations on where savings can be made in the short-term (e.g. by turning equipment off instead of leaving on stand-by) and long-term (e.g. by installation of double glazing or loft insulation).

Utility companies provide free simple domestic energy audits based on annual usage, number of rooms and number of radiators, provide an overall **energy label** for the property then give suggestions on how savings can be made. However, for a more comprehensive energy audit, the auditor must visit the property concerned.

Energy crops
Crops grown for the sole purpose of fuel use when mature. The **short rotation coppice** is the best example of cultivation of such crops. As with any fuel, energy crops vary and fluctuate in price. In the UK at present, about £40 per tonne of dry wood, adjusted down propor-

tionately for any remaining moisture content, is expected. **Miscanthus** and **switchgrass** are also energy crops of increasing importance.

Energy Institute (EI)

Founded in 2003 from a merger between the Institute of Energy and the Institute of Petroleum, the Energy Institute is a Royal Charter professional body and learned society. It is a financially independent organisation that develops and promotes knowledge to enhance good practice; support individual development; act as an honest broker in determining best practice (particularly in regard to industry operations through strong relations with regulators, industry and government); and facilitate debate and provide access to the science on energy-related topics. The EI is licensed by the Engineering Council (UK) to offer Chartered, Incorporated and Engineering Technician status to engineers and Chartered Scientist status to scientists working in the energy sector. The EI publishes industry standards, safety codes, guidance notes and technical publications.

Energy label

Information attached to electrical appliances (and other electrical devices such as light bulbs) to indicate to the consumer how much energy the appliance consumes. Energy labels allow the consumer to compare products on their energy efficiency and make a more realistic assessment of cost based on usage and not simply purchase price. In the European Union, most white goods must clearly display an energy label when offered for sale. The label displays the energy efficiency of the appliance in classes from A to G, A being the most energy efficient, G the least. The label also shows other useful information to the consumer to help them choose between various models.

Energy poverty

Energy poverty (or fuel poverty) is formally defined as a household spending more than 10% of its income on fuel use in order to heat the home to an adequate standard of warmth (defined by World Health Organisation to be 21°C in the main living room and 18°C in other occupied rooms). It does not take account of the amount that a household actually spends on fuel, nor the amount available for the household to spend on fuel after other costs have been met. Two million households in the UK in 2003 were thought to be in fuel poverty and thus had difficulty keeping their homes warm at an acceptable cost.

Energy-Return-on-Energy-Invested (EROEI)

Self-explanatory term for fuel production. Use of a particular fuel is usually viable only if it has an EROEI ratio of greater than one. EROEI figures abound on the internet but are not always clearly defined. In some EROEI figures, exploration has been factored in. Occasionally, EROEI figures are for exploration only. Care is therefore required in use of such figures. The EROEI of a fuel is obviously not a hard number. The EROEI for crude oil can vary from 20 for onshore oil in a shallow well to 5 for deep-sea oil. The value for natural gas is about the same as that for crude oil from the same source. Onshore, once the gas pressure inside the well drops and a 'nodding donkey' has to be installed, the EROEI goes down. For gasoline, the EROEI value depends on the EROEI of the parent crude and on how close the refining facilities are to the oil field. A widely accepted figure in 2006 for the EROEI of gasoline, production and refining energy costs both considered, was about 10. In 1950, it was 100 and in 1970 it was 25. Gasoline derived not from crude oil but from shale or tar sands tends to have lower EROEI values. Other typical values at the time of writing include 20 for hard coal and 10 for lignites. Again, these have to be understood as being values within a range: the actual EROEI value for a particular coal depends on factors including the depth of the deposit. The EROEI value for **ethanol** from corn is being quite vigorously debated at present: the Energy Bulletin for October 2006 gives a value of 1.3.

Use of a fuel of EROEI less than unity might be validated if the fuel would otherwise require professional disposal and the cost of such disposal is greater than losses incurred in using the fuel, or if use of such a fuel resulted in CO_2 reductions and losses were lower than such reductions expressed as 'CO_2 credits'. Household waste might conceivably fall in the former category and low-value biomass fuels such as **rice husks** in the latter.

Enhanced oil recovery (EOR)

Term applied to the pumping of gas into an oil well, on- or offshore, to increase its production. 60 million **bbl** of oil have been gained in this way in the **Magnus field** in the North Sea. **Oil India Ltd** have also used EOR techniques to very good effect. EOR can be applied not only to fields which are becoming depleted, but also to previously untapped fields where oil removal is made difficult by folds in the geological structure as at the **Idd El Shargi North Dome field**.

Enköping, Sweden
Site of **combined heat and power (CHP)** using both fuel wood (in the form of willow chips) and waste wood. The capacity is 22 MW of power and 55 MW of heat.

Enslow Fork mine, PA
The largest underground coalmine in the US. The coal is bituminous in rank and 10.2 Mt of it were produced in 2004. There are plans within the next few years to extend it to two additional seams. Almost as productive is the nearby Bailey Mine. Both are operated by **CONSOL Energy**.

Enthalpy
Function of state of a substance defined by:

$$H = U + PV$$

where H (J) is the enthalpy, U (J) is the **internal energy**, P (N m^{-2}) is the pressure and V (m^3) is the volume. The quantity is less fundamental than the **internal energy**, which follows from the First Law of Thermodynamics, but is preferred in most engineering applications including fuel technology for two reasons.

Firstly, if gas flow is under consideration as in applications including a turbine or a nozzle, the flow work is incorporated into the enthalpy and need not be considered separately. The flow work is not of any interest, so for it to have been deducted from the work which appears in a thermodynamic analysis is helpful. Secondly, the enthalpy change under constant pressure conditions is the heat transferred. Practical combustion systems, including grate combustion of coal, demonstrate approximately constant pressure.

Entropy
A thermodynamic quantity denoted by symbol S or s if on the kg basis. The Second Law of Thermodynamics states that the total entropy—that of system plus surroundings—increases as a result of any irreversible change. The Third Law of Thermodynamics states that all substances in their pure crystalline states have the same entropy at absolute zero of temperature. The importance of entropy in fuel technology is that thermodynamic cycles including the **Rankine cycle** are expressed in temperature-entropy diagrams. The units of entropy are J K^{-1}, hence the product $T \Delta S$ has units J (or J kg^{-1} if s is used). Heat transferred during such cycles in the limit where they are working reversibly is therefore calculable this way.

Environmental Protection Agency (EPA)

The EPA, sometimes referred to as the US EPA, is an agency of the federal government of the United States charged with protecting human health and with safeguarding the natural environment: air, water, and land. Established in 1970 by President Nixon, the EPA now employs over 18 000 people. Its work is to develop and enforce regulations; offer financial assistance for education; provide research grants and graduate fellowships; support environmental education projects that enhance the public's awareness, knowledge, and skills to make informed decisions affecting environmental quality; offer information for state and local governments and small businesses on financing environmental services and projects; and perform environmental research.

Epithermal neutron reactors

Nuclear reactors use either thermal neutrons, which are in thermal equilibrium with their environment, or fast neutrons which retain the energy with which they are released from a nucleus, there being no attempt to control this energy in reactor operation. The intermediate case is epithermal neutrons which are more energetic than thermal neutrons but have energies in a specified range, achieved by using a composite moderating material.

More relevant to medicine than to energy production, the epithermal neutron is now being used in a small number of nuclear reactors for power generation, at least on a test basis. Such applications are rare, however. More typically, either thermal reactors or **fast reactor**s are used in new plant for the generation of electricity from nuclear fuels.

Equatorial Guinea, hydrocarbon reserves of

This country in West Africa has major oil reserves although they are as yet largely untapped. Current production of oil is of the order 100 000 **bbl** per day from the offshore Zafiro field, and there is gas production at an offshore field off the island of Bioko, a province of Equatorial Guinea. A further discovery of **non-associated gas** off Bioko has been made at the time of writing by the Houston based Noble Energy. **Deep-sea oil and gas** is believed to be awaiting discovery in the Gulf of Guinea.

Erha field

Newly developed offshore field in Nigerian waters producing around 150 000 **bbl** per day by the end of 2006 as well as significant amounts of **associated gas**. The gas will, for the time being, be re-injected into the wells, which are at a depth of 3900 ft, making them the deepest production wells to have been drilled to date off Nigeria. The operator is ExxonMobil.

Eritrea, energy scene in

Exploration for oil off Eritrea, which has been taking place intermittently since the 1960s, has not been productive. The country consumes 6000 **bbl** per day of petroleum products, all of them imported in refined form. The only refinery in Eritrea has not been operational since 1997. Shell have a network of filling stations and also **liquefied petroleum gas** facilities. There is electricity generation at a mere 60 MW. A new oil-fired 88 MW facility was bombed before even coming into service although repairs have been effected. The transmission system is badly in need of renewal. Thanks to the aid agencies, there is significant power for schools and clinics from **photovoltaic cells**.

Escuinla, Guatemala

The terminus of a **pipeline**, under construction, to convey natural gas from Mexico to Guatemala and, in due course, from Guatemala to other central American countries. The pipeline is almost 350 miles long. The group of countries collectively known as central America consume no natural gas at present, not even the **associated gas** from the oil in Guatemala, so the installation of this pipeline will bring natural gas to the region for the first time. Its capacity will be 1 million m^3 per day.

Eskom

South Africa's largest electricity producer. Its power stations, currently 24 in number, burn 112 Mt yr^{-1} of coal, all of it domestic. Eskom is the largest consumer of coal mined in South Africa: **Sasol** is the second largest. Eskom produced 32 220 MW in 2005, and aim to increase this by 3000–4000 MW every three to four years up to 2020. In these expansion proposals, coal alone features as the fuel.

Esterhazy, Saskatchewan

Methane occurs not only in coalmines but also in salt mines and other mines yielding inorganic substances. Esterhazy is the scene of a potash mine where in January 2006 there was a severe fire. Investigators believe that the fire was ignited by a cutting torch which was being used to remove bolts from a flange connected to some polyethylene pipe.

Esters, as model compounds for biodiesel assessment

In view of the enhancement of **biodiesel** performance by esterification it is inevitable that esters have become model compounds in the study of biodiesels. Such esters are usually in the C_8 to C_{18} range e.g. methyl oleate. A difficulty with comparisons of esters with biodiesels is that whereas an ester such as methyl oleate has a single-valued vapour pressure at any particular temperature, a biodiesel (or a mineral diesel) does not: the measured vapour pressure depends on the space into which the vapour expands and all that can be measured for a biodiesel is the analogue of the Reid Vapour Pressure for a petroleum fraction. There have been attempts to correlate ester and biodiesel vapour pressures using the Clausius-Clapeyron equation for each. This is unsound as the C–C equation is for two phases of a pure chemical substance and does not apply to a mixture of compounds such as a biodiesel.

Ethane

Organic compound formula C_2H_6, the second in the alkane series. A gas at ordinary temperatures, it burns with a **calorific value** of 65 MJ m^{-3}. It sometimes occurs as a minor constituent of natural gas. The critical temperature is such that it can be liquefied by application of pressure only at ordinary temperatures and stored as a liquid under its own highly super-atmospheric vapour pressure. Ethane in this form is susceptible to **Boiling Liquid Expanding Vapour Explosion** behaviour if catastrophically leaked. Ethane in liquid-vapour form is used as a refrigerant in the chemical processing industry.

Ethanol

An organic compound (formula C_2H_5OH), also known as ethyl alcohol. It burns with a heat of combustion of just under 30 MJ kg^{-1}. It is used in suitably adjusted spark-ignition engines either on its own or blended with gasoline. Since ethanol is obtainable by fermentation, its use as a fuel is prevalent in countries which are major sugar producers, including Brazil. It is also used as a petrol blend stock in the US, in proportions as high as 85% as in **E85**. US states producing ethanol include Minnesota and Iowa. An ethanol plant is under development at **Aurora, NC**.

Ethiopia, energy scene in

This country has large reserves of natural gas and associated liquids. The Calub and Hilala fields, for example, contain about 100 **bcm** of gas and 13 million **bbl** of NGL and condensate (that is, about 2% of the total weight due to the liquids). A number of US and Canadian companies are currently active in development but there is at the present time no commercial production of gas as facilities including pipelines are absent. Ethiopia imports 4 million bbl of petroleum liquids per year. Ethiopia is one of only a few African countries to be involved in **geothermal electricity**, at about 12 MW level and its neighbour **Djibouti** is also likely to introduce geothermal power in the near future.

Ethylene
Alkene, structural formula $CH_2=CH_2$ and a product of **cracking** of petroleum fractions whereupon it becomes a starting material for subsequent synthesis. World production of ethylene in 2003 was ≈ 105 Mt. The International Union of Pure and Applied Chemistry name for this compound is *ethene*.

Eunice, Louisiana
The scene in mid-2000 of a major train accident in which tank cars with chemical payloads including ethylene oxide, acrylic acid and toluene derivatives were derailed. In the emergency responses, the car carrying ethylene oxide was purged by flushing with nitrogen, the ethylene oxide then being diverted to a **flare**. The cars containing toluene derivatives were exploded under controlled conditions. Homes within a two-mile radius were evacuated, but there were no deaths or injuries.

Evaporation losses
Such losses from stored and transported hydrocarbons carry significant fiscal penalties. They occur in all liquid hydrocarbons, from crude oil to light distillates. Mitigation methods include an annulus of water around a cylindrical storage tank and sprayed water onto the outside surface of a tank. Use of tanks with reflective outside surfaces also helps reduce evaporative losses and losses are also less in the cooler months. In general, about 1% of a quantity of crude oil will be lost by evaporation during storage in a tank for a year. The losses for a stored gasoline are greater, approaching 1% per month depending on ambient temperature conditions and the extent to which preventative measures such as those briefly mentioned above have been applied.

Everett, MA
Scene of the oldest **liquefied natural gas** import facility in the US, having been in service since 1971. The facility at **Cove Point, MD,** has a capacity about 20 % larger than at Everett (as at Elba Island, GA).

The facility at Lake Charles, LA, has a capacity of the order of twice that of the Everett facility.

Excess air

Fuels are often burnt under fuel-lean rather than fuel-rich conditions, which means that excess air is required. The precise meaning of the term is explained using the example of the combustion of methane below:

> If methane is burnt with 20% excess air, how many kg of air will be required per kg methane?
>
> $$CH_4 + 2O_2 \rightarrow CO_2 + 2H_2O$$
>
> *Solution*
>
> O_2 requirement: 2 mol per mol methane
>
> accompanying N_2 (2×3.76) mol
>
> Total air per 16 g methane = 64 g O_2 + 211 g N_2 = 275 g
>
> per kg methane, air requirement = 17.2 kg air
>
> Therefore with 20% excess air, 20.6 kg air per kg methane is required.

Excess air, in solid, liquid or gaseous fuel combustion, lowers the adiabatic flame temperature although the extent to which it lowers the actual flame temperature is controllable by operating conditions. Excess air is favourable in reducing oxide of nitrogen and particulate emissions. It allows economical combustion by minimising the amount of unburnt fuel remaining, which is also beneficial to the environment in that unburnt hydrocarbons are a factor in the formation of photochemical smog.

Excess crude oil

The **OPEC** nations always produce oil at an excess, partly because of their tendency to exceed quotas for exports. World excess in crude oil is always attributable to the OPEC nations. The situation in late 2004 was that excess crude oil was being obtained at between half a million and a million **bbl** per day, all which originated from **Saudi Arabia**. The excess oil is therefore of the order of 1% of the world total production. By way of perspective, note that the **Strategic Petroleum Reserve** contains an amount equivalent to the world's excess production over a period of two to three years.

Exploration and production licences, UK waters

An exploration licence is granted for three years and entitles the oil/gas company holding it to search for oil and gas in any UK offshore area not covered by a production licence. A production licence is effectively an operation licence: once granted it applies only to a small area within which the holder can explore and produce, and it remains in force until after eventual decommissioning. As a result of the **Fallow initiative** some exploration licences have, on changing hands, been rewritten to relate to redevelopment rather than exploration.

Exxon Valdez

Oil tanker which was the origin of a major leak in Prince William Sound, Alaska, in 1989. Spillage amounted to 11 million **bbl**. Litigation continues. To date Exxon have paid $287 million in compensation to fishermen, $1 billion in State and Federal settlements as well as $2.2 billion for the cleaning up operations.

F

Fairly Baram platform

Unmanned offshore platform which receives oil and gas from wells both in Malaysian and Bruneian waters. It is the scene of recent development work in the use of the **Coriolis meter** to determine flow rates of oil through pipelines as part of custody transfer procedures.

Falkland Islands, oil prospects in

Oil is believed to exist in significant quantities in the waters off the Falkland Islands, there being arrangements in place for liaison between the government of the Islands and oil companies seeking exploration licences. A difficulty is the lack of infra-structure, supposing large amounts of oil were to be found.

Fallow initiative

Currently underway in the UK sector of the North Sea, the Fallow initiative is intended to extend the lifetime of the oil fields by redeveloping some which have been 'fallow' for a period. Advances in methods for **enhanced oil recovery** enabled the **Magnus field** in the North Sea to come back into production and there are plans that the **Argyll field** will do likewise. In what is being termed the Fallow initiative, several other fields in the North Sea which have not produced for some time are to be redeveloped.

    The Department of Trade and Industry have divided the targeted fallow fields into two groups: Class A, in which such redevelopment is being carried out with all reasonable diligence by the present licensee, and Class B where it is not. Reclassification from B to A is possible if within three months the licensee produces an acceptable action plan for redevelopment, otherwise the licence is made available for purchase by other operators. On acquisition by a new operator, the licence is redrafted so as to appertain to a partly depleted source.

Faroe Islands, energy scene in

These islands between the northwest of Scotland and Iceland, with a population less than 50 000, generate electricity at about 25 MW. Some of this is hydroelectric, the remainder from **gensets**. There are proposals for **marine wave energy** using an **oscillating water column (OWC)**. The same organisation which set up such a facility at **Islay** is installing another in the Faroes. Another development, which could have a major impact on life in the islands, is oil exploration in the sea around them. A number of major oil companies have been granted exploration licences.

Farragon field

Newly developed oil field in the UK sector of the North Sea, with recoverable reserves of 18 million **bbl**. BP have a 50% holding. At the end of 2006 production was 18 000 bbl per day.

Fast reactor

When U^{235} is used as a nuclear fuel, processes such as

$$U^{235} + n \rightarrow Ba^{141} + Kr^{92} + 3n$$

occur. U^{235} is of course the minor isotope in terms of abundance. U^{238} is predominant but not easily fissionable, so in terms of nuclear fuel usage it is no more than a waste product. If, however, some of the neutrons released in processes such as that above can be made to impinge on to U^{238} atoms these undergo a process converting them to plutonium in the isotopic form Pu^{239}, which is fissionable, being the primary constituent of **mixed oxide (MOX) fuel**. The process is:

$$U^{238} + n \rightarrow U^{239} \xrightarrow[\substack{t_{1/2} = 23.5 \text{ min}}]{\beta \text{ emission}} Np^{239} \xrightarrow[\substack{t_{1/2} = 2.35 \text{ days}}]{\beta \text{ emission}} Pu^{239}$$

where Np denotes the chemical element Neptunium and $t_{1/2}$ the half-life of the process. This is the principle of a fast reactor, and a requirement is that one neutron released from each U^{235} fission is captured by a U^{238}. Such a neutron will be a fast neutron, retaining the kinetic energy with which it was ejected from a U^{235} nucleus. According to how the reactor is configured, it is possible for the fissionable atoms to increase in number, in which case there is a net increase—breeding—in the quantity of fuel and the reactor is known as a *fast breeder reactor*. However, not all fast reactors are set up to breed. The first significant power generation from a fast reactor was at **Dounreay**. Proposals for a fast breeder reactor, to ease dependence on imports of uranium, are under consideration in Japan. There has been an experimental fast breeder at Joyo in Japan since the 1970s. Provision of fuel for use there, at **Tokai**, led to a fatal accident in 1999.

Fault

Faults are planar rock fractures which show evidence of relative movement. Large faults within the Earth's crust are the result of shear motion. Active fault zones cause most earthquakes through the release of energy during rapid slippage along the faults. The largest examples are at tectonic plate boundaries. Oil or gas may accumulate and form a reservoir in a trap formed by one or more faults.

> The most famous is the San Andreas fault which divides the Pacific Plate from the North American Plate, running for 800 miles through California. Earthquakes caused by movement along this fault line can affect the large population areas of San Francisco, the City of Los Angeles and San Diego. Since Los Angeles and San Francisco are on opposite sides of the fault, they are moving towards each other at an average rate of 0.6 cm per year.

Fawley refinery

On the English south coast, near Southampton, this is the largest refinery in the UK, processing 300 000 **bbl** of crude oil per day, and operating since 1921. The first major expansion was in 1951, when the number of motor vehicles in the UK was sharply increasing. There have been several expansions since, including a cracking facility to make feedstock for the polymer industry.

Fiddler's Ferry, Cheshire, UK

Power station, previously coal fired, currently being adapted to coal and **biomass** co-firing at a projected level of 2000 MW of electricity. It will be the largest such facility in Britain. The biomass component of the fuel will include **energy crops**. An important part of the adaptation will be improved milling. Milling designed for coal often performs poorly in coal-biomass co-milling, and milling systems especially developed will be retrofitted at Fiddler's Ferry.

Field separation facility
Plant designed to remove **condensate** from a natural gas stream. May be sited on a gas platform or on land.

Filanovsky field
Newly discovered oil field in the Caspian Sea, to be operated by **Lukoil**. It is believed to contain 600 million **bbl** of oil and 30 **bcm** of **associated gas**. The area in which the field was found had not been explored for oil and gas in pre-Perestroika Russia, as the water above it was a breeding ground for the giant beluga sturgeon, the source of black caviar.

Filter cake
Coal product high in ash-forming constituents, a product of **coal cleaning**. The cleaned coal has a reduced **ash** content, and leaves a residue known as filter cake which can be used as a fuel in applications including the raising of steam. The combustion behaviour of filter cake can be affected by the catalytic effects of the inorganic matter present.

Fire
The exothermic chemical reaction accompanied by intense heat released during a rapid oxidation of combustible material. It may be visible as a brilliant glow along with flames and may produce smoke. To be sustained, fires need sources of heat, oxygen and fuel, known as the fire triangle. If any of these are removed, the fire will be extinguished.

Firebox
The enclosure within a piece of combustion plant where the combustion reaction itself occurs. Its surfaces need to be capable of withstanding the high temperatures involved and are constructed of **refractory materials**.

Firedamp
Methane having entered a coal mine by release from a coal deposit in which it was previously trapped.

Fiscal horsepower

Horsepower is a pre-SI unit of power, convertible to watts. The fiscal horsepower of a vehicle is the actual horsepower adjusted by a factor which depends on the cylinder diameter and the swept volume. At one time in the UK and in many other countries, purchase tax on cars was on a scale according to the fiscal horsepower. In common parlance, the horsepower of a car was the fiscal horsepower, not the brake horsepower. It sometimes happened that the name of a particular model was simply the fiscal horsepower to the nearest whole number: a well-known example is the Morris Eight.

Fischer-Tropsch

Named after the two chemists who designed the process, Franz Fischer (1852–1932) and Hans Tropsch (1839–1935). The original driver for the process was the lack of petroleum in Germany; petroleum had displaced coal as the main source of fuel. While Germany had abundant coal reserves it lacked sources of petroleum. The Fischer-Tropsch process is used to convert coal or natural gas to liquid fuels, via a **synthesis gas** of hydrogen and carbon monoxide:

$$CH_4 + \tfrac{1}{2} O_2 \rightarrow 2H_2 + CO$$

$$(2n{+}1) H_2 + nCO \rightarrow C_nH_{2n+2} + nH_2O$$

The types of products produced depend on process temperature, type of catalyst, pressure and gas composition. For gasoline and light **olefins**, the high temperature Fischer-Tropsch process is used, at 350°C. For distillates and waxes, the low temperature process is used, at 250°C. Catalysts used are typically iron or cobalt based (the latter having a longer life but being more expensive).

Fish ladder

Device enabling fish to ascend a hydroelectric dam. It was observed in one of the very early hydroelectric installations in the western US that salmon were prevented by the presence of the dam from making the movements and migrations necessary to their life cycle. Accordingly, the fish ladder was developed and widely used. It works in the same way as a lock on a canal.

Fishing

Term applied in the drilling of oil wells for the removal of part of the drilling assembly which has become detached and lodged in the well, delaying further drilling. A common 'fish' is a tooth from a drilling

bit. Devices for fishing include magnets, suction devices and, in extreme cases, explosives.

Fixed carbon

The amount of non-volatile residue of organic matter remaining in a sample of coal, **coke**, or bituminous material after ash, moisture and volatile matter have been removed as part of the **proximate analysis**. It should be noted that fixed carbon includes other elements from non-volatile components.

Flame

An exothermic reaction propagating subsonically through a mixture of fuel and oxidant. The colour and temperature of the flame are dependent on the type of fuel involved in the combustion process. Not all flames are visible to the naked eye, for example, hydrogen. The oxidant does not necessarily have to contain oxygen; for example, hydrogen burning in chlorine produces a flame and gaseous hydrogen chloride (HCl).

Flame arrester

A solid surface which will absorb heat and, possibly, reactive chemical intermediates at incipient flashback at a **burner**, in particular that at a **flare**, and prevent its development.

Flame photometry

One of about four widely used instrumental methods of analysis for **sulphur dioxide**. The method requires a hydrogen flame and can be used as a single device or as a gas chromatograph detector. The device responds to sulphur, and the degree of response depends on the chemical environment of the sulphur admitted. This means that equal amounts of sulphur admitted as sulphur dioxide and, for example, as a thiol vapour would not give exactly the same response in a flame photometric detector. Amounts of sulphur to the order of a nanogram are detectable.

Flame speed

Also known as velocity of flame propagation or burning velocity, it is formally defined as the velocity, relative to the unburned gas, with which a suitably defined flame front travels along the normal to its surface. In other words, it is the speed at which an area of burning gas travels through a combustible gas mixture towards the unburned gas.

Flare

The scene of intentional burning of unwanted hydrocarbons at an offshore platform, refinery or petrochemical plant. The flame at a flare is non-premixed and therefore tends to be smoky. Heat-release rates at flares are high, often of the order of MW.

Flash point

Index of the storage and transportation safety of a flammable liquid. It is the minimum temperature of the bulk liquid at which there will be a flash if the vapour above the liquid is contacted by a flame. It can be determined in an open cup (OC) or closed cup (CC) apparatus. For any one liquid, the OC value will be a few degrees higher than the CC value. Procedures for determining flash points are given in standards issued by bodies such as ISO and **ASTM**.

Flash pyrolysis

Exploratory method of obtaining petroleum substitutes from coal. Coal particles are admitted to a **fluidised bed**, the atmosphere of which is inert, possibly composed of post-combustion gas from another process. Contact between the fluidised particles and the coal particles allows very effective heat transfer. The coal particles experience rapid heating rates which enhance the yield of liquid breakdown products. These can be refined into the equivalents of petroleum fractions.

Flexible-fuel vehicle (FFV)

A vehicle which can function on any blend of fuel in the composition range from pure gasoline to **E85**. Most of the manufacturers which supply cars to the US make FFV versions of at least some of their models available. When such a vehicle changes hands, it can be confirmed from the identification number that it is in fact FFV. FFVs are expected to find favour because of the resolve of the US to reduce crude oil imports from Islamic oil-producing countries.

Floating production, storage and offloading (FPSO)

A FPSO is a widely used device in the offshore oil and gas industry. It receives oil from the well by means of flexible risers, the oil initially entering tanks in the FPSO. This is followed by transfer of the oil to specially designed vessels called shuttle tankers which take it ashore. An FPSO can be moved from one scene of oil production to another, but its attachment to any one wellhead is a major undertaking. An FPSO so installed is therefore expected to remain in service on a long-term, but not necessarily permanent, basis. It is prevented from drifting by mooring.

There are about 15 FPSOs currently in service in the UK sector of the North Sea. The **Roncador field** is also the scene of significant FPSO activity. Production of any one FPSO is up to 200 000 **bbl** per day. An FPSO can be made with a suitably reinforced standard ship's hull with some of the other parts taken from redundant offshore plant.

Flotta
Orkney island off northeast Scotland and the site of a major oil terminal from the North Sea Fields.

Flow calorimeter
Device available in various forms, perhaps the most common of which is the Boys calorimeter. A fuel is burnt at atmospheric pressure and the **calorific value** of the fuel is determined from measurements of the temperature of water in a jacket enclosing the device. As the burning is at constant pressure, the calorific value is the **enthalpy** change (usual symbol ΔH).

Flue gas
Gaseous products of combustion generated in a furnace. These are not necessarily the gases that will be emitted to the atmosphere since many combustion processes include some form of flue gas clean-up to remove the oxides of nitrogen and sulphur.

Flue gas desulphurisation (FGD)
See **desulphurisation**

Flue gas recirculation
In a coal-fired power station, recirculation of 20–30% of the flue gas at 300–400°C back into the furnace or to the burners decreases the **flame** temperature and oxygen concentration and helps reduce the formation of **thermal NOx**. Flue gas recirculation is used in combination with **low NOx burners** and furnace air staging (see **staged combustion**).

Fluidised bed
Combustion technique particularly suited to low-grade solid fuels. A bed of hot sand or coal ash is fluidised by entry of air at sufficiently high velocity (exceeding the velocity at incipient fluidisation). Fuel particles are also admitted to the bed. They receive heat from the flu-

idised particles by contact i.e. by conduction, more powerful than from a gas by convection. There is significant activity worldwide in coal combustion by this method, also in fluidised bed combustion of less attractive fuels such as high-ash coals, **culm** and even spent shale, in which some of the carbonised kerogen remains.

Fluorine, oxides of

These include F_2O and F_2O_2 which can be used as oxidants in the propulsion of a **rocket** as an alternative to liquid oxygen. Hydrogen (H_2) as fuel with F_2O as oxidant is a very powerful combination. The performance in terms of thrust is exceeded when compared to using hydrogen with liquid oxygen as the oxidant.

Fly ash

The ash resulting from **combustion** of **coal** as **pulverised fuel (pf)**. The coal particles are initially only of the order of tens of microns in size, hence the ash particles are micron or even sub-micron in dimension. Fly ash is removed from the post-combustion gas by an **electrostatic precipitator**. Fly ash might subsequently be usable in the manufacture of concrete. Examples of fly ash vary in colour from grey through neutral to red-brown, depending on what inorganics are initially present in the coal. The whole point of pf combustion is that the use of fine particles affords a high degree of completeness of combustion, in contrast to grate combustion which is often significantly incomplete. Well-conducted pf combustion results in fly ash of extremely low carbon content.

Food, transportation costs of

Many of the foods we purchase have been flown from overseas destinations in order to reach our supermarkets in fresh condition. There have been estimates that for products having been transported 1500 miles—the distance from Athens to London—a calorie of energy in food has required 10 calories in aviation fuel to bring it to its place of sale. Of course, much fresh produce in the UK has travelled significantly further distances than that, possibly from South Africa or South America, in which case proportionately more than 10 calories of fuel have been expended per calorie of nutriment.

Forepeak, oil tanker

Progressively narrower part of the tanker's hull terminating at an apex. Newly designed tankers are required not to hold oil in this part of the vessel because of its vulnerability in a collision.

Forest thinnings

Forests sometimes have to be thinned i.e. some wood removed to make the forest less susceptible to fire in the warmer months. The thinnings usually go to a landfill, but are obviously a possible source of waste wood fuel. Little numerical information is available on the extent of its use as fuel, since the various countries tend to aggregate wood waste from all sources. There is, however, a record of activity in Boulder County, Colorado, whereby local forest thinnings are the sole fuel for a **combined heat and power** installation.

Forties field

The largest oil field in the UK sector of the North Sea, entering operation in November 1975 and peaking at half a million **bbl** per day in 1978. Current production is about 81 000 bbl per day, having been just over half that in 2003. One of the production platforms has a facility for obtaining **condensate** from the **associated gas**. When the Houston-based Apache took over the field from BP, they increased production by improving the infrastructure. **EOR** by means of gas injection has also been carried out.

Fossil city

A city that is heavily dependent on the use of fossil fuels to uphold its economy, infrastructure and transportation networks.

Fossil fuel

A fuel produced by the decomposition and physical and chemical alteration of plants and animals over geological time. The term is usually used to describe the three fuel groups: coal, oil and natural gas.

Fossil fuels, *per capita* consumption of

The value averaged over the whole world is about 1500 W. It varies widely from one region to another.

FPSO

See **Floating production, storage and offloading**

Forward contract

An agreement to purchase or sell a certain amount of energy or fuel, on a set date, at a predetermined price. Contracts are usually settled physically with, for example, electricity being bought on the chosen date at the previously agreed price, independent of the actual price on the day (see **futures**).

Foundation Coal

The fifth largest coal producer in the US. Foundation Coal has thirteen mines, some in West Virginia and some in Illinois. Production for 2004 was 61.4 Mt. Most of the coal is sold to electricity producers: seventy buy coal from Foundation. The proposed reduction in imported oil to the US will in some states be achieved by use the use of coal, either as such or in gasified form, and this will require strategic responses by the major coal companies.

Fourier, Joseph

French mathematician who developed the Laws for conduction heat transfer which bear his name and which feature in all calculations on conduction. Born in France in 1768, Fourier published his earliest work while at a Catholic Seminary which he entered aged 21, although he did not complete his studies there and enter the priesthood as originally planned. Instead, he returned as a teacher of mathematics to the school where he had been a pupil. He held a number of posts subsequent to that one and eventually became Professor of Analysis at the École Polytechnique. In 1798 he took active military rank as scientific advisor to Napoleon Bonaparte during his invasion of Egypt. On returning to France in 1801 he resumed his post at the École Polytechnique. Thermal scientists with whom Fourier had professional dealings include Biot and Carnot.

One of Fourier's contributions is particularly pertinent during these early years of the 21st century. In an essay published in 1824, Fourier asserted that gases in the atmosphere could raise the temperature of the earth. In this he was way ahead of his time, as the seminal work of Stefan on thermal radiation was not to come until the 1870s.

France, petroleum products in

France was an early entry into the oil business; in the 1870s there were French refineries processing crude oil imported from the US. By the 1920s, France was looking for oil in the Middle East. At that time, France had more privately owned vehicles per unit population than any other European country and the **Standard Oil** Company was well established there. France was of course occupied during World War II. Its refineries were returned to their owners after cessation of hostilities and several **grass roots refineries** were built shortly after the War. Because of a consciousness of the need for domestic oil, exploration was carried out with discoveries of oil at Parentis and at Lacq in the 1950s. Nevertheless, imports from the Middle East and Africa increased over the following decades. The early 21st century situation is that France produces crude oil at about 15 million **bbl** per year and imports about forty times that amount.

Franklin and Elgin fields

Discovered in 1986 and in 1991 respectively, Franklin field is in quadrant 29 of the North Sea and Elgin field in quadrant 22. These quadrants are in fact adjoining: 22 is directly north of 29. The fields are in shallow water but required drilling to a depth of 5300 m and are the deepest fields ever brought into production in the North Sea. The fields are now producing at up to 150 000 **bbl** per day of oil and 12 million m^3 per day of gas.

Franklin, Rosalind

Best known for her contribution to the discovery of the structure of DNA through her skills in crystallography, Rosalind Franklin (1920–58) made a significant contribution to the field of carbon science. Her work on graphitising and non-graphitising carbons in the early 1950s for BCURA paved the way for work on carbon fibres. Following Dr Franklin's premature death from ovarian cancer, Professor J D Bernal wrote in her obituary in The Times (page 3, 19th April 1958):

> "Dr Rosalind Franklin discovered in a series of beautifully executed researches the fundamental distinction between carbons that turned on heating into graphite and those that did not. Further, she related the difference to the chemical constitution of the molecules from which the carbon was made."

Freeport, TX

Scene of a new **liquefied natural gas** terminal about 70 miles south of Houston. ConocoPhillips, who have a 50% share in the facility, will receive 40 million m^3 per day which it will convert back to methane gas for its own use in the manufacture of **methanol** from methane. Dow will buy some of the remainder for the same purpose.

Fuel cell

Device by means of which electron exchange in fuel oxidation is manifest directly as an electrical current. In conventional power generation, the fuel is burnt to raise either steam (in the case of a **steam turbine**) or hot post-combustion gas (in the case of a gas turbine) which then undergoes a cycle having severe intrinsic limits on its efficiency even if working reversibly. Power generation with a steam turbine according to a **Rankine cycle** achieves efficiencies, expressed as heat in/ work out, of about 35% and the theoretical but unattainable upper bound on the efficiency is that of a Carnot cycle working between the same temperatures.

In a fuel cell, the two electrodes (anode and cathode) are immersed in an aqueous electrolytic medium which allows for ion transfer. The most common type of fuel cell uses hydrogen as fuel and molecular oxygen as oxidant, according to the overall reaction:

$$H_2 + \tfrac{1}{2}O_2 \rightarrow H_2O$$

The hydrogen, which is being oxidised, loses electrons at the anode according to:

$$H_2 \rightarrow 2H^+ + 2e^-$$

and the oxygen is reduced at the cathode according to:

$$\tfrac{1}{2}O_2 + 2H^+ + 2e^- \rightarrow H_2O$$

Hence electrons flow, via an external circuit, from anode to cathode thereby providing electrical current. The directness of the process makes for much greater efficiencies than are possible with thermodynamic cycles. A brief thermodynamic analysis is provided below.

The maximum work obtainable is the Gibbs free energy change of the process, having a negative sign since a reduction in Gibbs free energy denotes positive work done on the surroundings. In terms of the definition of the Gibbs free energy:

$$\Delta G = \Delta H - T\,\Delta S = -w_{max}$$

where G denotes Gibbs free energy, H enthalpy, S entropy, and T is the operating temperature of the cell. Using subscripts r for reactant and p for product:

$$-w_{max} = G_p - G_r$$

The heat out is clearly:

$$H_r - H_p$$

Hence the efficiency η is given by:

$$\eta = \frac{-(G_p - G_r)}{(H_r - H_p)} = \frac{\Delta G}{\Delta H}$$

For the reaction as given in the stoichiometric equation above, with reactants and products at 298 K (*SI Chemical Data Book*, Wiley),

---

$$\Delta H = -285 \text{ kJ mol}^{-1}$$
$$\Delta G = -237 \text{ kJ mol}^{-1}$$

Therefore

$$\eta = \frac{237}{285} = 0.83 \, (83\%)$$

This efficiency far exceeds that which any power *cycle* could achieve.

There are however secondary effects which reduce the actual efficiency from that calculated above, including incomplete reaction at the electrodes and poisoning of the electrodes by **carbon monoxide**. Also, electrical current in the external circuit leads to a voltage drop which causes a proportionate drop in ΔG, hence the work done drops significantly below w_{max}. Even so, in large-scale power plants using fuel cells, efficiencies of about 60% are realisable. Hydrogen/oxygen fuel cells are also being used on a trial basis to power public service vehicles in cities such as London and Madrid. The enhanced fuel economy of a fuel cell over a diesel cycle is shown dramatically in the **BOC GH2OST** vehicle. The main fuel cell types are: **alkaline, molten carbonate, phosphoric acid, polymer electrolyte membrane** and **solid oxide**.

Fuel NO$_x$

NO$_x$ (NO/NO$_2$) arising from reaction of the fuel content of a fuel with oxygen during the combustion process. Most fuel nitrogen goes to elemental nitrogen (N$_2$) during combustion but, depending on the combustion conditions, a small proportion will go to fuel NO$_x$ which, at sufficiently high combustion temperatures, will be accompanied by **thermal NO$_x$**. NO$_x$ is involved in photochemical smog formation and also in **acid rain**.

Fuel rods

U^{235} is the most common nuclear fuel for electricity generation and for such use will have been prepared as isotopically enriched uranium oxide pellets, installed in metal tubes typically 12 feet long. The fuel so installed constitutes a fuel rod. A bundle of such fuel rods comprises a fuel assembly and a nuclear power station might have several hundred fuel assemblies (the reactor) enabling the facility to operate without fuel renewal for approximately 1–2 years. Control of heat-release is by means of the control rods made of boron, which are capable of absorbing neutrons. To reduce the heat release rate these will be lowered into the reactor, neutron absorption having the effect of slowing down the net-releasing nuclear process. Clearly, withdrawal of the control rods will have the opposite effect where that is required. In addition to the control rods which are operated as necessary, there are additional rods which are programmed to enter the reactor if the nuclear process becomes too rapid. At UK nuclear power stations these cannot be overridden. A contributing factor to the Chernobyl accident in 1986 was that these emergency control rods had been overridden.

Heat from nuclear fuel is used to generate steam as in a thermal power station, the steam being used to drive a turbine.

Fujian terminal
Proposed **LNG** terminal for China with a capacity of 2.6 million tonne of **LNG** per year. Expected to begin operating in 2008, it will be China's second largest such terminal exporters to which will include British Gas and the **LNG** once 'regasified' will be used in power generation.

Furnace
A device used for heating for a particular purpose. For example, see **reverberatory furnace**.

Futures
An agreement to purchase or sell a certain amount of an asset (for example: energy or fuel) on a set date which may, or may not, result in a physical exchange. A futures contract gives the holder the right and the obligation to buy or sell; both parties must exercise the contract (buy or sell) on the settlement date. To leave the commitment, the holder of a futures position has to sell his long position or buy back his short position, effectively closing out the futures position and its contract obligations. Futures are exchanged trade derivatives, traded on the futures markets to transfer risk and generate profits. Whereas a **forward contract** is settled on the trade date at a previously agreed price, futures are settled at the price fixed on the last day of trading of the contract. Futures therefore carry less financial risk than forwards.

G

Galapagos Islands, oil spill at

These islands, straddling the equator off the coast of South America, are maintained largely as a habitat for wildlife. There was therefore some consternation when the 30-year old Japanese-built tanker vessel *Jessica*, carrying diesel and **bunker fuel**, ran aground at one of the islands in 2001. The Charles Darwin Research Station, which is concerned with wildlife protection, is itself situated on one of the Galapagos Islands and officers of the Station worked alongside the Ecuadorian army and the island authorities in dealing with the accident. Over the first few days after the spill, the vessel which had run aground was floated, its fuel tanks having first been sealed. The vessel was anchored and the remaining fuel contents—50 000 gallons—were removed. Dispersing agents were applied to oil which had leaked in the initial incident.

One year on, a report on the incident and its effects prepared by the Charles Darwin Research Station recorded that oil had spread from the island where the vessel had ran aground to several of the other islands. Many sea lions had to be treated for oil contact, and there was particular concern over the resident marine iguanas who are the sole survivors of the species. Fish and marine invertebrate populations were closely examined and monitored for a period after the spill.

Galp Energia Group

Formerly Petróleos de Portugal, the largest Portuguese oil concern. Primarily concerned in the refining of oil for distribution in Spain and Portugal, Galp has nevertheless expanded into exploration and production having, for example, a 9% holding in **Lucapa-1**. There is similar activity in Brazil and in Syria.

Galveston, TX, 1900 hurricane in

This hurricane in September 1900 was the first in the 20th Century to hit the Gulf coast, and claimed 6000 lives. It occurred only four months before the oil gush at the **Spindletop field**. During the very rapid expansion of the Texas oil industry which followed, oil companies established their business centres in Houston rather than in Galveston. It is possible that had the 1900 hurricane not occurred, Galveston would have had the place that Houston now has in world oil affairs, implying the indirect effect of this hurricane on the oil industry was immense.

Garyville LA

Scene of a refinery operated by Marathon, commencing production in 1976 and said to be the most recently built **grass roots refinery** in the US. Its capacity is 245 000 **bbl** per day making it currently the sixteenth largest refinery in the US. The recently expanded **Marathon Detroit Refinery** has a capacity of less than half that of the Garyville refinery.

Gas cap

The gas that is allowed to accumulate above the oil in a reservoir if pressure, temperature, and fluid characteristics permit.

Gas constant

Constant of proportionality in the equation of state for an **ideal gas**, usual symbol R. If the equation is set up in its most general form:

$$PV = nRT$$

where P = pressure (N m^{-2}), n = quantity in moles and T = absolute temperature. The gas constant R has value 8.314 J K^{-1} mol^{-1}. Because it applies to any gas, provided that the quantity in the ideal gas equation is in moles, this value is sometimes called the Universal gas constant.

In fuel technology, more than in most areas of science and engineering where the ideal gas equation finds application, there is a tendency to use a value of the gas constant specific to the gas, in which case a quantity m (kg) replaces the number of moles in the equation. Using air as an example, the molar mass is 0.0288 kg hence:

$$m = n \times 0.0288 \text{ kg}$$

and the ideal gas equation becomes:

$$PV = m \, \text{R}_{air} T$$

where R_{air} is the specific form of the gas constant for air and has the value:

$$\frac{8.314}{0.0288} = 289 \quad \text{J kg}^{-1} \text{ K}^{-1}$$

Gas oil

Once a synonym for diesel for the reason that diesel was sometimes converted by **steam reforming** to **town gas**. This is now an obsolescent technology but there is linguistic interest in the term in that is has survived in languages other than English to mean diesel in the conventional sense, for example in Italian and Spanish. The word diesel with the same meaning as in the English language exists in countries including Belgium, Denmark, Germany, Holland, Sweden, Israel and Switzerland.

Gas-to-liquid (GTL) fuels

These are made from natural gas and are intended as substitutes for diesel. They have negligible sulphur and very low aromatic content. The latter quality results in a low propensity to particulate formation on burning. Current applications of GTL fuels include a number of London buses.

There are plants for the manufacture of GTL fuels in countries including Malaysia (actually at **Bintulu**), South Africa and the US. A sum of about $20 billion has recently been committed for the development of a GTL plant in Qatar. The target production is 3×10^5 **bbl** per day in 2011.

Gasification

The thermochemical process of converting a carbonaceous solid or liquid fuel (**biomass**, **coal**, **coke** or **petroleum** oils) into a gaseous fuel ($CO + H_2$).

Gasohol

Term coined and used about 35 years ago in the American Midwest for a blend of gasoline and 10% **ethanol** for use in spark-ignition engines. Nowadays the petrol pump will simply display a notice that what is on offer is 10% alcohol.

A major gasohol plant is proposed for Thailand in association with Alfa Laval.

Gasoline, cost of

In the US, about 15% of the cost of gasoline is attributable to the refining process. Crude oil accounts for about 43% of the cost, Federal and State taxes about 30% and marketing and distribution the balance. Distribution costs vary since 50% of the gasoline used in the US is refined at Gulf coast locations and therefore has to be distributed over long distances. State taxes vary from 6 cents to 39 cents per gallon (in December 2005).

Gasoline in the US, Middle East content of

The US is now a net importer of oil, receiving from other countries 13.2 million **bbl** per day, 2.7 million bbl of which originate in Saudi Arabia, Kuwait and Iraq. All of the 150 refineries (approximately) of varying capacity in the US take at least some Middle East oil so there is no gas station which can claim to provide the product with no Middle East content although, of course, any one delivery might fortuitously be free of any material from the Middle East.

Gasoline prices, variations in

The price of gasoline depends on the price of crude oil and on the qualities of the gasoline, including its octane rating and presence or absence of lead. It depends more strongly on taxes imposed. In a country which is a federation of states or provinces there can be state/provincial tax on gasoline on top of the federal tax. These factors have to be considered when any comparative interpretation is made of prices. Prices below have been converted to US dollars.

The countries of the European Union sell gasoline to the motorist at prices in the approximate range $1.20 to $1.70 per litre. In the US regional variations are very wide but a price in the neighbourhood of 55 cents per litre would be typical. In Canada one would expect to pay 65–70 cents per litre, which also applies in Australia and New Zealand.

In 2005 there were major rises in the price of crude oil. Factors responsible included the death that year of King Faisal of Saudi Arabia and a sequence of refinery fires in the Gulf Coast states of the US. At the time of going to press crude oil prices are $60 per **bbl**. At times during the previous eighteen months they had been approaching $70 amid much international consternation.

Gasoline substitutes, octane ratings of

As examples, **methanol** has an octane rating of 108 and methane an octane rating of 120. Each of these is better than the octane ratings of even the most expensive conventional gasolines.

Gasoline yield from crude oil

The total refining capacity of the USA is 17 million **bbl** per day, which yields 7.48 million bbl of gasoline, a 44% yield. This is partly straight-run distillate but also material from the same crude chemically adapted for blending with the gasoline fraction e.g. reformed naphtha.

Gasonol

Blend of gasoline and 20% **ethanol** for use in spark-ignition engines, once widely used in the Philippines.

Gaza Strip, electricity supply to

By formal agreement, Israel supplies electricity both to the Gaza Strip (administered by the Palestine Authority) and to the part of Palestine known as the West Bank. However, two municipalities within the West Bank—Nabulus and Janin—have small plants of their own for electricity generation.

Gengibre well

Exploration well in Angolan waters drilled by Marathon Oil, having given highly promising results. The well is in 1700 metres of water, somewhat deeper than those in the **Girrasol field**, and was drilled to ≈4400 m depth. Marathon have made altogether nine deep water oil discoveries off Angola.

Genset

Generating set, a device usually working on a diesel cycle whereby output is passed along to an alternator and AC produced, often at tens of kW although sometimes up to about 2 MW, with 50 Hz or 60 Hz frequency as required. Clearly a genset has the advantage of mobility and enables power to be generated in areas remote from the grid or where there has been failure of supply from the grid. In some applications, including welding and lighting, a genset might be incorporated even where a reliable power supply is available. The genset is a more established device than the **microturbine** with which it will be in competition for some applications, power production being in the same range for the two types of device. Major manufacturers of gensets include Volvo, the Mitsubishi Corporation and Iveco.

Geothermal electricity

This utilises heat from within the Earth as a basis for driving a turbine and hence making electrical power. The power plant may take one of a number of forms, depending on the temperature at which heat is extracted from the Earth.

If water occurs inside the Earth at temperatures in excess of about 150°C (as it does at locations such as northern California), on release into the atmosphere this will form steam which can be used to drive a turbine in a conventional way. Such an installation at Puna, Hawaii, produces power at 25 MW.

Alternatively, if the water resource does not have such a high temperature, its heat can be applied to a liquid with a lower boiling point than water (often isobutane) and the vapour produced used to drive the turbine. At the conclusion of the thermodynamic cycle, the vapour is not, for obvious reasons, simply released at a cooling tower but is recycled for further heat exchange and turbine passage. Such a system of geothermal electricity production, involving two fluids, is called a binary plant. The Casa Diablo facility in California consists of three such plants, and has a combined capacity of 27 MW.

Electricity generation is not of course the only application of geothermal resources. Very often the previously subterranean hot water is used simply to provide heat for buildings or for recreational facilities including swimming pools and, at **Copahue, Argentina** for example, for the clearance of snow.

Germany, natural gas production by

Germany has 3×10^{11} m^3 of recoverable natural gas reserves, most of the gas fields being onshore. One gas field in the German sector of the North Sea also produces **condensate**. The former German Democratic Republic (East Germany) relied on **town gas** made from **lignite**. There is also significant activity in **coal bed methane**, and it is proposed to install a power generating plant at the site of several disused mines.

Germany, wind power in

Currently seen as the world leader in the generation of power by **wind turbine**, Germany is expanding in this activity by 750 MW by turbine installations at two newly selected locations. Turbines at these sites will be assessed for future offshore use of the same designs.

Germany's total generating capacity by this means is currently 4.5 GW, almost twice that of the USA.

Ghawar field, Saudi Arabia

The world's largest oil field, discovered in 1948 and productive since 1951. Current production is around 5 million **bbl** per day. The field was developed in phases over the decades, some new wells being sunk in a previously untapped region of the field as recently as the mid 90s. The Ghawar field produces about 6% of world's crude oil. The **associated gas** at Ghawar accounts for a third of **Saudi Arabia**'s known natural gas reserves.

Ghislenghein

Located about 30 km from Brussels, Belgium, and the scene of a natural gas accident in July 2004 in which there were about 20 deaths and over 100 injuries. The accident, involving two successive fireballs, occurred during a repair to a pipeline. Police and fire fighters were among the dead. The pipeline being repaired carries gas to northern France.

Giant field

According to an arbitrary definition by the American Association of Petroleum Geologists, a giant field is one which contains, on discovery, 500 million **bbl** or more of crude oil. The term is encountered fairly frequently in reports of exploration and appraisal.

Gibraltar, energy and power in

Gibraltar is self-sufficient in terms of electricity, generating about 10 MW from imported **fossil fuels**. There is excess capacity for contingencies as there is no provision for import of electricity from Spain. Shell, using the name Shell Gibraltar, supply the only two filling stations in Gibraltar and also the small airport. In Gibraltar Bay, Shell operate a floating facility to provide **bunker fuel**, and vessels from several countries avail themselves of this. Gibraltar is therefore strictly speaking an exporter of oil.

Gilberton, PA

Scene of a proposed plant for making diesel fuel from local coal. Shell and **Sasol** will provide know-how. The Governor of Pennsylvania has backed the proposal by undertaking that the State and its associated trucking facility will buy 40% of the plant's products. Target production is 5000 **bbl** per day.

Gillette, WY
Scene of a major deposit of **sub-bituminous coal** with an annual production of 4.8 Mt. Operated by the Wyodak Resources Development Corporation, the mine could continue at its current level of production until at least 2060.

Girrasol field
Deep-sea oil field off the coast of Angola, discovered by Elf Exploration in 1996. Wells have been drilled 1400 m below the sea and production facilities are being installed. These include risers to take the oil to a **floating production, storage and offloading**.

Gleaner Oil and Gas
In the UK there are a number of independent companies who buy Shell products for distribution and sale in the remote and lightly populated areas. An example is Gleaner Oil and Gas, which supplies Shell gasoline and heating fuel to the northern Highlands district of Scotland. Its head office is in Elgin, roughly midway between Aberdeen and Inverness. The gasoline is sold at Gleaner filling stations, not at Shell stations. Other such companies include Swan Petroleum, which supplies Shell products to parts of Wales and the border counties and also Manx Petroleum which provides a similar service for the Isle of Man. Gleaner Oil and Gas has another responsibility in addition to supplying Shell gasoline to the Highlands: it also supplies Shell **liquefied petroleum gas** to the whole of Scotland.

Global warming
Human-induced climate change is threatening to impose very significant shifts in temperatures, rainfall, weather extremes and sea levels in this century and those that follow. The principal cause is the increase in carbon dioxide levels in the atmosphere due mainly to humanity's growing use of **fossil fuels**, which trap more solar warmth.

The Greenland and Antarctic ice sheets are stratified layers providing a continuous record over half a million years. For at least the last 30 000 years, annual layers can be identified. The isotopic composition of the ice enables a reliable estimate to be made of the air temperature at the time the ice was formed. Gas bubbles within the ice are thought to provide a reliable record of the concentration of carbon dioxide in the global atmosphere shortly after the ice formed. Records in cores taken from the Vostok ice sheet in Antarctica indicate carbon dioxide concentration and temperature over the last 400 000 years.

Changes in the carbon dioxide concentration in the atmosphere have been closely correlated with changes in temperature. A rise in temperature causes carbon dioxide to be released from the oceans, thus increasing the concentration in the atmosphere. However, an increased concentration of carbon dioxide causes temperature to rise.

Warm tropical oceans release carbon dioxide, while cold, polar oceans absorb carbon dioxide. If changes in solar radiation raise the temperature of the Earth's surface, there is a net release of carbon dioxide from the oceans, the concentration in the atmosphere increases, and further warming occurs because of the greenhouse effect; that in turn releases more carbon dioxide, and so on in a positive feedback. The reverse process occurs if changes in solar radiation result in a cooling of the Earth's surface. Complicating factors include the expansion and contraction of ice sheets (which absorb less solar radiation) and the factors governing the rate at which carbon moves into and out of the deep oceans.

Human-induced climate change is inevitable. The global challenge is to halt the steady rise in the concentrations of carbon dioxide and other greenhouse gases, limiting further change and reducing the risks of catastrophic alterations in climate.

Gluckauf
The first oil tanker vessel built in Britain, completed in 1886.

Gokasho Bay, Japan
Scene of **marine wave energy** presently at a set level of 110 kW. The **oscillating wave column** there is known as the 'Mighty Whale' and has a subsidiary purpose: it acts as a wave breaker, to the benefit of fisheries behind it.

Gonfreville Refinery
The largest refinery in France, operated by Total (strictly TotalElfFina). Its capacity is 328 000 **bbl** per day. In addition to refining there is also a **combined heat and power** facility in which Texaco have a 50% share.

Gorgon field
One of several newly developed gas fields off northwest Australia. It is believed to contain about a quarter of a trillion cubic metres of

natural gas, and the intention is that gas from the field will be converted to **liquefied natural gas** for export to countries including South Korea.

Grangemouth
A centre of hydrocarbon processing, with inventory brought from the North Shore fields, situated close to the coast of Fife, Scotland. Activities include refining, polymerisation and the production of **natural gas liquids**.

Granite Falls, Minnesota
The scene of electricity production at a rate as high as 75 MW using alfalfa stems as fuel. Alfalfa waste is used as fuel in other parts of the US including Idaho.

Grape residues
These have some potential as fuel and there is significant activity in Spain where 10^5 tonnes of such residue is produced each year, capable of releasing on combustion about 1 PJ (PetaJoule, 10^{15} J). Olive residues also find fuel use in Spain.

Grass roots refinery
A refinery built *ab initio* on a site previously devoid of hydrocarbon plant. It contrasts with expansion or renovation of an existing refinery, however extensive. The most recently constructed grass roots refinery in the US is in **Garyville, LA**, although it is in fact 30 years old. The refinery proposed by **Arizona Clean Fuels Yuma** will, on completion, become the most recently built grass roots refinery in the US and will therefore close the rather remarkable three-decade gap.

Greasecar system
Means by which a conventional diesel vehicle can be made to run on **straight vegetable oil (SVO)**. The vehicle needs to have two fuel tanks and the engine is always started up on mineral diesel from one tank. Heat from the engine is transferred to the SVO which is in a second tank, and a temperature is reached at which the SVO has a viscosity suitable for the fuel injection system of the engine. At that stage, the diesel supply is shut off and the vehicle runs on SVO.

There have in the US been many success stories with cars adapted to run on SVO (see http://www.greasecar.com/profiles_list.cfm) and there is also activity in European countries including Spain. An inconvenience entailed is that if the vehicle is stopped and the engine allowed to cool to ambient temperature the SVO resident in the engine will need to be flushed out before the next start-up.

Greater Sunrise field

Offshore gas field believed to contain 300 million **bbl** of condensate also. Australia and Indonesia will each work the field, although a final settlement has not yet been reached.

Greece, fuel resources of

All coal originating in Greece is **lignite**, annual production being about 60 Mt most of which is used to generate electricity. There are six lignite-fired power stations with a combined generating capacity of 4850 MW and two further 550 MW power stations are planned. As at the **Latrobe Valley** in Australia, the power stations are situated close to the mines.

Greece's reserves of oil and gas are extremely limited, being confined to one offshore field and an extension off the island of Thasos, producing less than 2 million **bbl** of crude oil per year.

Green energy

The use of **renewable energy** sources such as solar, wind and wave power instead of **fossil fuels** such as coal, oil and natural gas.

Greenhouse gas emissions, reductions of in the UK

Latest available figures for 2005 show greenhouse gas emissions (comprising six gases) falling to 656.2 Mt from the 1990 base year value of 770.3. Carbon dioxide emissions (which account for 84% of the total) were 554.2 Mt, down from 592.1 over the same period. The reduction in overall gases is encouraging auguring very well for the strategies being implemented.

Greenland, oil reserves in waters off

Oil companies including ExxonMobil and Chevron have applied for exploration licences off Greenland. Previously there had been six exploration licences in these waters, none of them at all productive. There have however been observed seepages and geological surveys indicate the areas of most promise, so subsequent exploration wells will not be **wildcats**. A hazard with offshore activity off Greenland is possible iceberg impact on to platforms.

Gresford, Denbighshire
The scene of a coal mining disaster in 1934 in which 266 lives were lost, one of the three mining accidents in 20th century Britain in which the death toll was in three figures. The others were at **Senghenydd, Glamorganshire** and at **Hulton, Lancashire.**

Greymouth Petroleum Company
Greymouth is a small town on the west coast on the South Island of New Zealand. The Greymouth Petroleum Company, set up in 2002, produces oil and gas from three fields: Kaimiro, Ngatoro and Turangi. The Company also holds a number of exploration licences within New Zealand. The Greymouth Petroleum Company's most recent (April 2006) discovery was natural gas at Turangi, believed to be accompanied by condensate in amounts of the order of 5 million barrels.

Groensund Strait
Stretch of sea between Denmark and Germany where 764 000 gallons of oil were released when a tanker collided with a freighter in 2001. The tanker was in fact double-hulled and was carrying 9.7 million **bbl** of oil.

Groningen
Site of a major gas field in the Netherlands, with $\approx 2 \times 10^{12}\, \text{m}^3$ of reserves at the time of discovery. Groningen currently accounts for two-thirds of the Netherlands' total off- and onshore gas reserves. The former are in the Dutch Sector of the North Sea.

Grozny
Location of an oil field in the Caucasus region, with a position comparable to that of Baku in the regional oil industry. Oil was discovered there in 1833. Hitler's failure to gain control of the oil reserves at Grozny was a factor in his eventual defeat. Both Baku and Grozny, although continually producing oil, diminished dismally in productivity between World War II and Perestroika.

GTL fuels
See **gas-to-liquid fuels**

Guajira Peninsula
Region of Colombia about 60 miles from the Caribbean coast, having vast reserves of **bituminous coal** and the source of the very significant exports of coal from Colombia.

Guam, oil recycling in

This US Territory imports liquid fuels from Singapore and economises by using waste oil where possible in such applications as steam raising. The penalty for the use of such low-quality fuel is heavy combustion plant maintenance costs. There is a new scheme whereby a private company will take the waste oil, refine it and return it for use as **bunker fuel**. As would be expected the scale is small, about 48 000 **bbl** of the refined fuel per year for which the refining costs will be $US 1.8 million, meaning that a good general purpose liquid fuel is being provided locally at a price of $US 37.5 per bbl.

Guangdong

Province of southern China and the scene of **hydroelectricity**. This has not however been at a level to make the province self-sufficient in terms of power and there have been exports from other parts of mainland China and also from Hong Kong. In 2003–04, demand from the hydroelectric facility at Guangdong so exceeded supply that rationing of electricity became necessary.

Guatemala, oil production in

This country has 526 million **bbl** of oil reserves. There has been much foreign investment in developing the oil fields and production is now in the region of 20 000 bbl per day. There are storage areas for crude oil but as yet only one refinery, with a capacity of 2000 bbl per day. Recently, an exploration licence has been granted to the UK based Taghmen Energy who will investigate a 78 000 acre site in the northwest of the country.

Gullfaks field

Oil field in the Norwegian sector on the North Sea operated by Statoil, a mature field (production commenced 1986). There is also **associated gas**. Three production platforms are in service at the field. Gullfaks is currently the scene of improved production by **underbalanced drilling**. Several new wells have by this method been drilled and Statoil predict another 20 years of production at this field which was previously believed to be approaching depletion. Consequently application of **underbalanced drilling** to other Norwegian offshore fields is expected.

Gujurat, India

Scene of construction of a **grass roots refinery**, the completion of which was delayed by cyclone damage in 1998. Once the facility is fully operational, output is expected to be about 10 Mt yr^{-1} of fractionated product, making the refinery the second largest in the sub-continent. The boost to the availability of motor fuels in India, which the refinery will eventually bring about, will make room for 17 000 new retail outlets for such fuels. These will be phased in over a period of a few years by the Indian company Essar Oil, who will be the owners of the refinery.

Gulf of Mexico, daily oil production
Production in the Gulf of Mexico is 1.5 million **bbl** per day. There are over 3000 platforms in the Gulf, the total capital value of which exceeds $US 100 billion. They are vulnerable by being in a region susceptible to hurricanes.

Gulf of Mexico, first recorded hurricane in
This was in 1527, at which time Spanish conquistadors were exploring the Gulf of Mexico. Five vessels bearing conquistadors encountered a hurricane and were, in the words of one historian, 'tossed like driftwood'.

Gypsum, blending of with coal ash
Coal ash can be blended with gypsum to produce useful materials. Ash from the coal gasifier at **Wabash River, Indiana** is so used in the manufacture of wallboard. Note that the ash from Wabash is from a gasifier bed, not from a **pulverised fuel** burner, and so will consist of sizeable lumps.

Hadong Power Station

Power generation in South Korea near Pusan, conventionally designed and operated using steam in a **Rankine cycle**. The fuel for steam raising is imported bituminous coal and there are six 500 MW turbines giving a nameplate capacity of 3000 MW. The Korean company Hyundai is licensed to manufacture locally Babcock steam raising plant.

Hammer, Armand

American oil magnate, latterly president of Occidental. He had a very unusual background as a businessman, having qualified in medicine but never practising, going instead into marketing and export of pharmaceuticals and later asbestos mining and pencil manufacture. His association with Occidental began in 1956 when he invested in two wells in California each of which was unexpectedly productive.

Occidental's international activity included oil and gas production in the North Sea, where their platform Piper Alpha was destroyed in 1988 with the loss of 167 lives. Armand Hammer, by then aged 90, was one of a team of Occidental management who fielded questions from the media after the accident. In the widely used BBC video on the Piper Alpha accident the statements he made in the company's defence are criticised by the commentator.

Hammerfest liquefied natural gas (LNG)

There has been recent activity in **LNG** at the Snøvit, Askeladd and Albatross gas fields, north of the Norwegian town of Hammerfest within the Arctic circle. The fields were discovered over 20 years ago and the gas from them contains appreciable amounts of condensate. Sub-sea production plant will release gas from about 20 wells and pipelines will take the gas to a **barge** berthed at the small island of Melkøya 160 km away. The sub-sea production operations will be monitored from the island. The barge is equipped with facilities to convert the gas to LNG after removal of the condensate which will also take place on the barge. The barge was built in Spain and towed to Melkøya. Production of 4.3 Mt yr^{-1} of LNG is expected by late 2007.

Hamza field

Oil field in Jordan, found by a foreign operator in 1982. Production began in 1985. It is the only oil field in Jordan ever to have produced anything, although current oil recovery from the field is negligible. Nevertheless, it contains about a third of a million **bbl** and the infrastructure remains in place, so it is possible that the field has a future.

Harbin

Chinese city of population approximately equal to that of London i.e. 7.2 million, threatened by leakage of benzene from a petrochemical plant about 200 miles away. Benzene released during an explosion at the plant entered the nearby river and was carried to Harbin, causing contamination of the water supply to homes, instigating panic buying of bottled water and foodstuffs.

Hard coal

Generic term for **anthracite** and/or **bituminous coal**. A **brown coal** is soft by comparison.

Hard coals, crushing energy required for

Usually in the range 800–1800 kJ per tonne of coal, depending on the **Hardgrove index** of the coal and the efficiency of the crushing device.

Hardgrove index

Empirical method of determining the grindability of a coal. Coal is admitted to the Hardgrove apparatus which is manually operated, each turn of the handle supplying a precisely known amount of grinding energy. In standard applications, 60 turns are applied to coal previously reduced in particle size to the range 0.6–1.2 mm. The Hardgrove index is calculated from knowledge of what proportion of the coal after grinding will pass through a mesh of size 74 mm. The higher the Hardgrove index the more grindable the fuel. The method is used not only on coal as mined but also on coal products including **briquettes**.

Hashima Island
Japanese territory in the East China Sea, off the Nagasaki Prefecture. Bituminous coal suitable for coking was discovered there in 1887. In 1890, Hashima was acquired by Mitsubishi which had, in 1881, purchased the mine at Takashima, an island also off the Nagasaki Prefecture. This was a period when industrial and military expansion in Japan were rapid, and by the approach of World War I production at Hashima was 150 000 tonnes per year and the population of the island, comprising solely mine employees and their families, was 3000. Annual production had increased to 410 000 tonnes by the time of Pearl Harbour. Hashima was one of a number of Japanese coal mines where, during World War II, forced labour from Korea and China was used. Coal production continued at Hashima until 1974.

Hazira
Coastal location in the Gujurat region of India and the scene of a new **liquefied natural gas (LNG)** terminal with a capacity of 5 Mt yr^{-1}. The first delivery occurred in April 2004, the LNG having come from the new liquefaction facility off the northwest shelf of Australia. Shell has a significant interest in the terminal at Hazira and Shell's LNG tanker Gemmata, capacity 135 000 m^3, made the initial delivery in 2004. The vessel itself was newly commissioned at the time.

Heat recovery
A process that utilises waste heat from a plant for a useful purpose, for example, to generate electricity or district heating (see **combined heat and power**).

Heat storage capsules
An exploratory device whereby wax encapsulated in a tightly sealed plastic outer coating is incorporated into building structures. As the building receives heat, from the sun or from artificial sources, the wax takes up some of the heat by melting and such heat is therefore unavailable as sensible heat to raise the temperature of the building. As the heat is withdrawn the wax melts and releases its latent heat which

then becomes sensible heat and partly compensates for the withdrawal of heating. The net result is therefore a smoother temperature history of the building over a 24-hour period. In effect, the thermal mass of the building is raised and a small building can be made to behave like a much larger one in terms of heat gain and loss.

Heavy crude

A crude oil having an unusually high density. There is a good deal of heavy crude in Venezuela, which can be blended with a conventional crude before export. Further upgrading by hydrogenation gives a modified crude oil with a density of about 32 degrees API and very suitable for refining. In reporting and statistics a heavy crude, once upgraded, is regarded as a conventional crude.

The distinction between a heavy crude and **bitumen** is not clear-cut and the latter is sometimes referred to as 'extra heavy crude'.

Helicopters, fuel for

Helicopters use the same fuel as fixed-wing jet aircraft, that is, suitably specified kerosene.

Helium density

Density of a coal measured by displacement of helium. Helium is capable of entering all of the pores of the coal down to the smallest in the micropore range and accompanying sorption of helium on to the pore walls is negligible. The density so measured is that of the solid, the voids being excluded.

Hemp (*Cannabis sativa*)

Plant crop which has for many centuries provided a basis for the manufacture of such things as coarse fabric (the word canvas is a contraction of cannabis) and rope. Hemp seed oil is seen as having promise as a **biodiesel**. Such oil is rich in aliphatic acids, esterification of which is required before the oil can be substituted for mineral diesel in compression ignition engines. Such esterification is straightforward to carry out with **methanol** as a reagent.

Because of possible misuse of hemp to make a narcotic drug, its cultivation requires a licence in many parts of the world including the EU and the US.

Henry Hub

Close to the town of Erath Louisiana, the scene of vast natural gas handling. About half of the natural gas from the Gulf of Mexico and in the onshore Texas and Louisiana fields passes close to Henry Hub, where over fifteen major pipelines are interconnected. Accordingly, Henry Hub is the counterpart for natural gas pricing of **Big Sandy River** in **bituminous coal** pricing and the 'Henry Hub spot price' for gas the counterpart of the **CAPP** price for coal. This is for gas distant from the wellhead with **natural gas liquids** and condensate removed. Generally in natural gas pricing, the basis is not mass or volume of gas but amount of heat released on combustion in units of millions of **British Thermal Unit (BTU)**. The Henry Hub spot price for natural gas currently is $US 7.51 per million BTU. (The price varies considerably; one week previously it was $US8.91, but 12 months previously just 20 cents lower.)

Hibernia
The name of a recently developed oil field off the Newfoundland coast. A shallow field, it is believed to contain 600 billion **bbl** of recoverable oil. Impact between the production platform and an iceberg had to be considered in the risk assessment. A direct hit from a one-Mt iceberg has an estimated frequency of 2×10^{-3} per year, or once every 500 years.

High-grading
The practice of developing and extracting resources of the highest quality and lowest cost (see **low-grading**).

Hindenburg
At 19:30 on 6th May 1937, the Hindenburg dirigible, the largest aircraft ever to fly, was destroyed by fire and explosions as it was about to land in Lakehurst, New Jersey. 62 passengers survived but 35 lost their lives. The Hindenburg was nearing its landing site during an electric storm. According to observers on the ground, the dirigible began to drift past its landing position, and after a brief delay, the ship started to vent hydrogen into what was highly charged outside air: a combination of factors that could prompt severe corona activity on any airship. An eyewitness reported seeing a blue glow of electrical activity above the Hindenburg before the fire started, indicative of the extremely high temperatures typical of a corona discharge. As the crew attempted to bring it back on course, control of the ship was lost, the tail touched the ground, and the stern burst into flames. Passengers who were afraid the ship might explode, jumped to their deaths. The burns and other

injuries were a result of the diesel fuel fire, not from hydrogen. Most of the passengers, who waited for the airship to land, walked safely away from the accident.

The Hindenburg is often cited by members of the general public as an argument for not moving towards a **hydrogen economy**. Until recently, it was thought this tragedy had been caused by hydrogen, the highly flammable gas used to inflate the skin of the ship. However, historical photographs show red-hot flames. Hydrogen burns invisibly. Also, no one reported smelling garlic, the scent which had been added to the hydrogen to help detect a leak. The mystery of the Hindenburg was solved by a team of NASA scientists. By using infrared spectrographs and a scanning electron microscope they were able to discover the chemical makeup of the organic compounds and elements present in the fabric of the dirigible's skin. They found that the Hindenburg was covered with a cotton fabric that had been treated with a doping compound to protect and strengthen it. However, this compound contained a cellulose acetate or nitrate, effectively gunpowder! Aluminium powder, which is used in rocket fuel, was also identified. The outside structure was wooden and the inside skeleton was duralumin coated with lacquer. The combination was flammable and deadly. Sadly, the Hindenburg disaster ended the use of airships for passenger voyages.

Hinzhou, China
Scene of a major refinery which handles 12 Mt yr^{-1} of hydrocarbon. Some of its products, in particular naphtha, are diverted to a new petrochemicals facility in Shanghai of which Shell own 50%. The petrochemicals facility will also receive condensate from one or more of China's non-associated gas fields.

Holford Gas Storage facility, Cheshire, UK
Underground storage facility for natural gas currently under construction. It will be owned and operated by the German company E.ON and is expected to be in full use by 2010, at which stage its total capacity will be 165 million m^3 of natural gas. Cheshire was once the principal salt producing region of Britain, having several large salt mines. The gas storage facility will be in eight salt caverns created by pumping water to the level where the storage enclosures are required, so dissolving the salt. The resulting brine will be pumped to the surface. In Germany itself, the combined capacity of all of the underground storage facilities for natural gas is 19 **bcm**.

Hong Kong, electricity supply in

In this very vibrant part of the world, power generation is entirely in the private sector. Companies making electricity are not franchisees: any organisation which can satisfy regulatory requirements can enter the electricity market. There is thermal generation in Hong Kong using coal, natural gas and oil and also some generation with nuclear fuels. Total power supply is about 5000 MW. At present about 7% of the power generated in Hong Kong is exported to mainland China, although over the next few years this could be reversed. There is minor activity in power generation from renewable sources including **wind turbines** and **photovoltaic cells**.

In the associated territory of Macau there is a power facility in Macau City (using **gensets**) and two in outlying islands (one using gensets and the other having two gas turbines and a **steam turbine**). The total capacity of these three facilities is 488 MW which is not quite sufficient to supply the territory. Import at an average of 29 MW supplements the power produced internally.

Hongai

Urban region of Vietnam previously known as Haiphong, having **anthracite** reserves. Much of the anthracite mined is exported to western Europe. One Dutch company sells anthracite from this source after processing under the brand name Hongai.

Honkeiko, China

The scene of what is recorded as the worst ever coalmining accident, when 1549 lives were lost. The accident occurred in 1942 and the cause was a coal dust explosion. Standards of safety in Chinese coalmines are still dismally low. For example, in 2002 18 miners were killed in a coalmine explosion in southern China and in mid-2005 24 miners were killed in an explosion in the Yuzhou mine in the Henan Province.

Honshu Island

One of the islands comprising the nation of Japan. The sea close to it is the scene of the only offshore hydrocarbon production in Japan. Honshu is also the scene of **geothermal electricity generation**.

Hoolamite

Reagent used to detect **carbon monoxide** in the atmosphere. The reactive component is iodine pentoxide which on contact with carbon monoxide changes colour.

'Hot pot'

Colloquial term for hot potassium hydroxide solution, once used at offshore production platforms to remove carbon dioxide from any gases re-injected into the well.

Hubbert Peak

Hypothesis formulated in the 1950s by Dr M King Hubbert regarding the depletion of any finite resource such as crude oil, where the maximum in production corresponds to 50% depletion. If this is so, it follows that matters presently are at such a peak and that 50% of the world's crude oil reserves have been used up, only about a century after significant crude oil usage began. Advocates of renewable fuels such as **biodiesel** frequently invoke the Hubbert peak in support. There are however counterviews and the US Geological Survey expects that we will not arrive at the peak before the mid 2030s in the US.

Hugoton gas field

The largest natural gas field in North America. Most of it is in southwest Kansas although it extends over the Oklahoma and Colorado state borders. Productive since 1928, it yielded 0.7 **bcm** of gas in 2004. Further exploration with a view to extending the field is being carried out by Occidental. Southwest Kansas has a number of such fields, including the Bradshaw, Bryerly and Greenwood fields, which jointly meet 92% of the natural gas needs of the State. Some of this gas is **associated gas**: oil in annual quantities of 8–10 million **bbl** accompanies it.

Huitengxile wind farm

Situated in inner Mongolia, this facility is capable of producing power at a rate of 26 MW using 22 turbines. Under the **emissions trading** scheme, the credits realised by the wind farm have been purchased on behalf of the Dutch government.

Hulton, Lancashire

Scene of a coal mining accident in 1910 in which 344 men were killed, the second worst pit accident in the UK in the 20th century. One factor in its initiation is believed to have been a faulty lamp, which would have provided an ignition source for **firedamp** that had leaked into the mine atmosphere.

Humble Oil and Refining Company

Formed in Texas in the late 19th Century and once active in the

Wilmington field. Humble (pronounced Umble) Oil was in fact the forerunner to Exxon.

Humidity

The ratio of the pressure of the vapour in a vapour-gas mixture to the saturated pressure of the vapour at the same temperature. Conventionally, the vapour is water and the gas atmospheric air. However, in chemical engineering the definition is broadened and one can quite correctly talk of the humidity of, for example, a benzene-nitrogen mixture, water being totally absent.

Hundred-year wind

Term used in risk assessment at offshore installations in the Gulf of Mexico, where there are records of wind speeds going further back than 1945 when offshore hydrocarbon in the Gulf began. The value currently stands at 40 m s^{-1} (90 mph). When assessing risk to a platform, the frequency with which the platform will experience this wind speed is assigned a value of 0.01 per year. The hundred-year wind speed is much less than the highest speed observed during Hurricane Katrina, which was about 175 mph.

Hurghada

The location of an oil field in Egypt, productive from 1915 until the mid-1980s. Oil was discovered there in 1913 and the field was initially operated by the Anglo Egyptian Oil Company (AEO) in which Shell and the Anglo-Persian Oil Company (later BP) each had a 50 % holding. Later AEO was nationalised and, as the El-Nasr Petroleum Company, continued operations at Hurghada. Other assets of the El-Nasr Petroleum Company include a refinery near Suez which has a capacity of 150 000 barrels per day. Since 1998 there has been a sizeable wind farm at Hurghada.

Hurricane Audrey

Occurring in 1957, one of the first hurricanes to strike the Gulf Coast after the commencement of offshore oil and gas production there. With the exception of the two of the **Scorpio** design, all of the offshore installations were damaged.

Hurricane Ivan

This hurricane in the second half of 2004 necessitated the evacuation of 3000 persons from offshore installations in the Gulf of Mexico and reduced total oil production in the Gulf by over 60%. Tankers taking oil from Venezuela were also delayed for several days. The **Strategic Petroleum Reserve** was drawn on to mitigate the effects of the Hurricane on gasoline supply.

Hurricane Katrina
Occurring in August 2005 and causing many deaths in the southeast coastal states of the US, this hurricane also necessitated the evacuation of many oil and gas platforms in the Gulf of Mexico. Some of these platforms were not able to function until major repairs had been carried out, while others were deemed a total loss. There were also effects on the downstream side of the industry including the devastation of several Louisiana refineries, some of which were completely submerged in water. Refineries in the Gulf States and elsewhere, which previously received oil from the affected platforms, had to look elsewhere. Plant sited close to the Mississippi River exploded, resulting in the pollution of the river.

The **Henry Hub** spot price as a benchmark for natural gas pricing was suspended as a result of the hurricane, which happened only a matter of months after **Hurricane Ivan**.

Hurricane Rita
Occurring only a few weeks after **Hurricane Katrina**, Rita did not bring about the same degree of destruction or impact in the Gulf States. It nevertheless caused serious damage to three refineries, including one at **Port Arthur, TX**.

Hybrid
Term relating to a hybrid energy system where two or more forms of energy or power are combined to provide an energy service. It is also used as a term for hybrid electric vehicle, of which there are two types. The first uses an internal combustion engine to power a generator which charges a battery, which is then used to power electric motors. The second is powered from both an internal combustion engine and an electric motor. The **Supermajors** are all developing hybrid vehicles using closely guarded technology.

Hydrazine
Chemical compound formula N_2H_4, used as a fuel in **rocket** propulsion. Suitable oxidants are oxygen, fluorine or ClF_3.

Hydrocarbon pipeline, total length of in the world

The total length is c. 5 million miles. One third is in North America.

Hydrochloroflourocarbons (HCFCs)

Compounds containing carbon, hydrogen, chlorine and fluorine. Although ozone depleting, they are less destructive than **chlorofluorocarbons**. See entry on chlorofluorocarbons for the effects of HCFCs on ozone depletion.

Hydrocracking

Also known as destructive hydrogenation, this process is a high-pressure, catalytic, refinery process that involves **cracking** of heavy petroleum fractions in the presence of an excess of hydrogen which suppresses the formation of tars and **coke** and also improves the crackability of **polyaromatic hydrocarbons**. The process produces high conversion rates of gasoline and diesel fuels or can supply feeds for other refinery processes. Tungsten sulphide is used as the catalyst for vapour phase hydrocracking and an iron catalyst or bauxite is used for hydrocracking in the liquid phase.

Hydrofluorocarbons (HFCs)

Compounds containing carbon, hydrogen and fluorine. HFC compounds have been developed as alternatives to **chlorofluorocarbons** and **hydrochlorofluorocarbons** which are both ozone-depleting due to the presence of chlorine. HFCs are still considered to be greenhouse gases and to contribute to **global warming**.

Hydroelectricity

The generation of electricity by passing water through a turbine. Its analysis according to the First Law of Thermodynamics is extremely simple, as the numerical example below demonstrates.

> Imagine that it is required to generate electricity at 100 MW by installing a turbine at a depth z of 80 m below the surface of a lake. The overall efficiency of the turbine is 75%. At what rate will water need to flow through it?
>
> The power required is 100 MW therefore the rate of supply of mechanical energy is:
>
> $$\frac{100}{0.75} = 133 \text{ MW}$$
>
> At turbine exit, the speed of water flow can be taken to be

negligible, and the potential energy is also zero as the turbine depth is the level at which, for the purposes of thermodynamic analysis, the vertical co-ordinate is zero. All of the mechanical energy at the exit is 'pressure energy'. Letting the pressure at this point be P N m^{-2} and using the symbol σ for the density of water: copied version first

$$\frac{P}{\sigma} = \frac{\sigma zg}{\sigma} = zg = 80\text{m} \times 9.81\text{ms}^{-2} = 785\text{m}^2\text{s}^{-2}\left(\equiv \text{Jkg}^{-1}\right)$$

The rate of water passage is then:

$$\frac{133000000\text{Js}^{-1}}{785\text{Jkg}^{-1}} = 170 \text{ tonne s}^{-1}$$

Major users of hydroelectric power include Scotland and Australia. There is also major activity in Fiji. This is an isolated island country with negligible indigenous fuel resources and hydroelectric generation is especially suited to such. The scene of the generation is the mountains of Vanua Levu, the second largest of the Fiji islands. Much expansion of hydroelectric power in China is currently underway. Chile has long had a dependence on hydroelectric power: nearly three-quarters of its electricity was produced by these means in the mid -1990s. There is expansion of hydroelectric activity in Chile at places including the **Bio Bio River.**

Hydroelectricity is obviously advantageous in that it involves no release of gaseous pollutants of any kind into the atmosphere. On the other hand, its reliability can be affected by a spate of dry weather, leading to a reduction in water resources as happened at **Guangdong** recently.

Hydrogen, elemental

This exists as the bimolecular H_2 and is combustible according to:

$$H_2 + \tfrac{1}{2}O_2 \rightarrow H_2O$$

for which the heat of reaction is 286 kJ mol^{-1}, equivalent to 11.9 MJ m^3 (288 K, 1 bar) of hydrogen reacted. Hydrogen is a constituent of many manufactured gases including **producer gas**, **retort coal gas** and coke oven gas. It is also a **devolatilisation** product of many coals. Hydrogen alone is an increasingly important fuel not least because its combustion obviously does not involve any CO_2 emission. It can be prepared from synthesis gas, the key reaction being:

$$CO + H_2O \rightarrow CO_2 + H_2$$

requiring a temperature in the region of 400°C and a catalyst, possibly iron oxide or chromium oxide. There is currently much activity in the production of renewable hydrogen from synthesis gas made from the action of steam on **biomass**. Hydrogen can also be prepared by the electrolysis of water. Use of elemental hydrogen in transport is chiefly in **fuel cells**. Hydrogen dispensers for hydrogen-powered vehicles, ranging in size from buses to small private cars, are being installed at filling stations in countries including the US and Japan.

Hydrogen, production of by algae

The algal species *Chlamydomonas reinhardtii*, when deprived of sulphur, transforms its metabolism to anaerobically produce adenosine triphosphate (ATP) with hydrogen gas H_2 as a side product. This may be a promising renewable source of hydrogen for fuel use, and research continues.

Hydrogen economy

A hypothetical future economy in which energy is stored as hydrogen for transport applications and to provide the daily peak demand reserve for the electrical grid.

Hydrogen peroxide, accidents due to

Recent years have seen two serious accidents involving the explosion of hydrogen peroxide, which, like high explosives such as TNT, requires no external oxidant in order to sustain an exothermic reaction. In 1999, a road tanker on the Metropolitan Expressway in Tokyo which was bearing hydrogen peroxide exploded and there were several fatalities. The tank had previously been used to carry copper chloride solution and had not been cleaned before admittance of the peroxide. Reaction between the copper compound and the peroxide is believed to have caused the explosion. In 2005 a tanker carrying hydrogen peroxide exploded on the M25, motorway near London. There were no fatalities: the driver of the tanker was treated for injuries. Closure of the motorway in both directions for a number of hours resulted.

Hydrotreating

A catalytic process in an oil refinery in which **hydrogen** is contacted with petroleum product streams from light **naphtha**s to lubricating oils to remove impurities such as sulphur, nitrogen, oxides, halides and trace metals. The catalyst used is cobalt/molybdenum based. Hydrotreating is milder than the **hydrocracking** process.

I

Idd El Shargi North Dome field

Field in shallow water off Qatar where the stratigraphy is such that oil is trapped in uneven structures and difficult to remove. Occidental have applied **enhanced oil recovery (EOR)** techniques there and production began in 1999 at a rate of 2500 **bbl** per day. The oil, once at the well head, is transferred 15 miles by pipeline to the production platform at Idd El Shargi South Dome field, which is itself the scene of EOR.

Ideal gas

The most fundamental definition of an ideal gas is that it is one for which the **internal energy** is a function of the temperature only. Another definition is that a gas is ideal if it obeys both Boyle's Law and Charles' Law, in which case the equation of state incorporating the **gas constant** applies. Difficulties occur when the specific heat at constant pressure and constant volume (symbols C_p and C_v respectively, units J mol^{-1} K^{-1}) calculated for an ideal gas according to the fundamental thermodynamic definition given above are tested against values for real gases, as conformity can only be expected if the real gas is monatomic e.g. helium. There is a view that the term 'perfect gas' applies to a monatomic ideal gas, but more generally the term perfect gas is used synonymously with 'ideal gas' without the qualification that the gas is monatomic.

Ideal gas enthalpies

An arbitrary measure of the **enthalpy** of air. In fact, all measures of enthalpies of a substance have to be on an arbitrary basis since there is no one temperature at which all substances have the same **enthalpy** as there is, by reason of the Third Law of Thermodynamics, for entropies. Calorimetry determines enthalpy changes, in going from reactants to products, and not actual enthalpies. For many engineering purposes the **enthalpy** of dry air is set to zero at 0 K where:

$$h_a = cT$$

where h_a is the specific enthalpy of air in J kg^{-1}, c the heat capacity in J kg^{-1} K^{-1} and T the temperature in Kelvin as indicated. One will find enthalpies for air so calculated in reference works including *Perry's Chemical Engineer's Handbook*, values of c of about 1000 J kg^{-1} K^{-1}, varying slowly with temperature, being used. Hence the ideal gas enthalpy of air at 300 K is given in the Handbook as 300.2 kJ kg^{-1}.

IEA

See **International Energy Agency**

I G Farben

German chemical giant whose activities included the production of aviation fuel from **braunkohle** during World War II.

Illawarra

Region of New South Wales, Australia, south of the state capital Sydney, where there is coal production and also integrated coal mining and steel manufacture. When an industry dominates a particular place over the order of a couple of generations or more it starts to determine the culture of that place and the traits of its inhabitants. The author D H Lawrence was the son of a coal miner in the English midlands. When D H Lawrence lived temporarily in Australia in the 1920s he based himself in the mining town of Thirroul, in the Illawarra region.

Illipe meal

Seeds from the Borneo tallow nut tree (*Shorea stenoptera*) which occurs in Malaysia and other eastern countries and yields on crushing a semi-sold material widely used in cosmetics as the basis of creams. The solid residue remaining (meal) retains some of this material and is a good biomass fuel. Some Illipe meal is in fact imported into the UK as a **biomass** fuel.

Imam Khomeni Port
At the southwestern tip of Iran, and poised for expansion which will make it Iran's major port for **hydrocarbons** and also for other exports. Its proximity to the major refinery at **Abadan** was one incentive for the development of the Port.

Imhotep-1X
Recently developed gas/condensate well in the western desert of Egypt, yielding about ≈ 1 million m^3 of gas per day. Also in the western desert region is the Mihos-1X well, about 25% more productive of gas than Imhotep-1X. The combined condensate yield is 2330 **bbl** per day. The wells are in the Matruh field, discovered as **wildcats** by **Shell** in 1991.

Incineration
Incineration is the thermal destruction of waste. Modern incineration systems use high temperatures, controlled air, and excellent mixing to change the chemical, physical, or biological character or composition of waste materials and are equipped with state-of-the-art air pollution control devices to capture particulate and gaseous emission contaminates.

Incineration can be adapted to the destruction of a wide variety of wastes. This includes, but is not limited to, household wastes, often referred to as municipal wastes, industrial wastes, medical wastes, sewage, and the hazardous wastes (liquids, tars, sludges, solids and vent fumes) generated by industry. The major benefit of incineration is that the process destroys most of the waste rather than just disposing of or storing it.

Waste incineration involves the application of combustion processes under controlled conditions to convert waste materials to inert mineral ash and gases. The '3 Ts' of combustion (temperature, turbulence, and residence time) must be present along with sufficient oxygen for the reaction to occur.

Depending upon the physical and chemical characteristics of the waste and the handling they require, different incinerator designs are applied. Solids, sludges and tars are incinerated in fixed-hearth and rotary kiln incinerators. Liquids may also be burned in these systems and used as support fuel. In many plants where liquids are the primary wastes, liquid injection incinerators are used. Boilers, process furnaces, cement kilns, and lightweight aggregate kilns also use the energy available from liquid wastes and burn liquid wastes as well as the fossil fuels (natural gas and oil). Moving or reciprocating grates are commonly used for municipal solid waste incineration.

There are still many health concerns connected with incineration systems, especially for people living near incinerators. However, the stringent regulations ensure that the design, operation, testing and maintenance of these systems provide maximum safety and minimum risk to the surrounding area and population.

India, shale reserves of

In the Assam (also the location of the refinery at **Digboi** for conventional oil) and the Arunchal Pradesh regions of India there are vast reserves of shale and the additional benefit of ample water at the site of the deposits. There is also the space to do something constructive with the spent shale, such as build a hydroelectric dam with it. There is therefore much interest in developing these reserves and investors are sought.

An interesting point is that the proven amounts of shale in India exceed the minimum reserves required for admittance of the country to **OPEC**, although production would have to reach a minimum level to meet the criteria for full membership. Indonesia was faced with possible reduction from full member to observer because of a fall in production in recent months. OPEC will accept shale as reserves for the purposes of eligibility for membership, but will not accept **natural gas liquids** or **condensate**.

India, wind power in

A capacity of just over 6 GW in late 2006, making India the fourth largest generator of wind power. Approximately 95% of installed capacity is in seven states.

Indiana, use of landfill gas in

This state of the US has seven power plants which use **landfill gas (LFG)** as fuel, producing in total 22 MW of electricity. The most recent plant to come into operation is in White County, with a capacity of 3 MW. Local sources of LFG are sufficient for this level of generation to be sustained on a long-term basis.

Indonesia, geothermal power generation in

This country has major geothermal resources and at present, in spite of internal difficulties which have impeded development, it has a **geothermal electricity** capacity of 862 MW which is about a quarter of that of the USA. This is divided between installations at several areas including Salak (55 MW), Kamojang (330 MW) and Dieng (60 MW). Smaller pilot schemes are producing in the region of 2–5 MW and expansion of these is expected. The Indonesian power supplier **Perusahaan Listrik Negara** has geothermal as well as hydroelectric and conventional thermal generation in its portfolio.

Inertinite
A **maceral** group derived from plant material that has been strongly altered and degraded in the peat stage of coal formation. For example, fossil charcoal is inertinite maceral. The inertinite macerals are distinguished by their relative reflectances and structures.

Injection well
A well in which fluids are injected rather than extracted to increase pressure and stimulate production. Water is the preferred fluid as it can help push oil towards the production well.

Insurance premiums for offshore platforms
In the Gulf Coast annual premiums are about 0.75% of the value of the platform. Hence a platform worth a $1 billion (typical value) would cost $7.5 million to insure for a year. The recent hurricanes are expected to lead to increases in these figures, perhaps by as much as 150%. Lloyds are well represented in the Gulf and have a major centre in Houston.

Interconnector
The single cable electricity supply link between the United Kingdom and France, intended for two-way use to top-up either the French or British electricity network, with a capacity of 2000 MW. Laid in 1986, the interconnector has only been used one-way from France to Britain. A second interconnector was constructed in 1998 between Bacton and Zeebrugge for the transportation of natural gas between Europe and the UK (or *vice versa*). The UK import capacity will be 23.5 **bcm** per year from 2007.

Interior West, atmospheric pollution in
A number of power facilities in this region of the US have been identified as major releasers of greenhouse gases and atmospheric pollutants. They are the Intermountain Power Project UT, Four Corners and San Juan NM, Mohave and Reid Gardner NV and Navajo Generating Station AZ. Power from these plants is exported to California,

and they emit oxides of sulphur and nitrogen much more abundantly than California's own power plants because of the more stringent emission limits in California than in UT, NM, NV and AZ. There have been not altogether unreasonable suggestions that these four states are in effect becoming dumping areas for California's atmospheric pollution. The four power facilities identified in this entry release 67 Mt of carbon dioxide annually. The carbon dioxide emissions from the Intermountain Power Project UT significantly exceed the total emissions from all of the motor vehicles registered in Utah.

Internal energy

Function of state, symbol U or u if on a kg basis, units J and J kg^{-1} respectively. The First Law of Thermodynamics states that for any process, the balance of heat transferred and work done is the change in internal energy.

International Energy Agency

A Paris-based intergovernmental organisation founded by the Organisation for Economic Co-operation and Development (OECD) in 1974 following the **oil crisis**. It is dedicated to preventing disruptions in the supply of oil, as well as acting as an information source on statistics about the international oil market and other energy sectors. It has a secondary role in promoting and developing alternative energy sources, rational energy policies, and multinational energy technology cooperation. It holds a combined stockpile of 4 billion barrels, 1.4 billion of which governments control for emergency use. Much of the oil is held in the form of petroleum products which need no further processing. The one sector it does not study in detail is nuclear fission, which is covered by the International Atomic Energy Agency.

InterOil

Canadian company with interests primarily in Papua New Guinea. Operations include exploration, refining and retail distribution of petroleum products. It operates alongside other much bigger concerns such as ExxonMobil. InterOil's refinery near the national capital Port Moresby began production in 2004 and receives local crude. Previously crude from Papua New Guinea had to be sent to Australia or Singapore for refining.

Ipati Block

Scene of a major natural gas discovery in Bolivia by Total. It is estimated to contain 300 **bcm** of natural gas. The Argentine company Tecpetrol have a 20% interest.

Iran, oil and gas reserves in
Huge amounts of crude oil exist in this country, where oil production began in 1913. Current annual production is in the neighbourhood of 10^9 **bbl**, and a further 70 years at that rate of production have been projected on the basis of the proven reserves. About three-quarters of Iranian oil is exported. Iranian oil has much **associated natural gas** and there are also non-associated gas fields, most notably the **South Pars field**. If and when the **Iran-India pipeline** comes to fruition, some of this gas will travel along that facility.

Iran, wind power in
One does not traditionally associate the **OPEC** countries with developments in renewables. However Iran, one of the earliest members of **OPEC**, raises several GW of power from wind farms. There are wind farms at seventeen locations in Iran, administered by the Iran Atomic Energy Organisation.

Iran-Armenia Gas Pipeline
This will carry gas from Iran to Armenia reducing dependence of the latter on supplies of gas from Russia. Expansion is planned to achieve an eventual doubling of the initial capacity. Neighbouring Turkey has also been adversely affected by the unreliability of natural gas supplies from Russia. The pipeline was due to enter service late in December 2006 but although the line was ready, tests had not been completed. It is planned to become operational in March 2007.

Iran-India Pipeline
The pipeline will, when fully complete, carry about 1.7 million m^3 of gas per day to India. The pipeline will pass through Pakistan, transit charges being payable, and it is possible that Pakistan will itself purchase some gas from the pipeline. Its length will be 2775 km. The US have discouraged India from the enterprise and offered instead to assist India in creating a new nuclear facility for power generation. Also, during the weeks of debate on the project, India has discovered an enormous offshore gas field at **Andhra Pradesh**.

Iraq, current position of in OPEC
Iraq is excluded from recent production figures for **OPEC** countries,

and no production quota has been assigned by OPEC to Iraq since 1998. A further difficulty is that the *ex officio* representative of a member country at an OPEC Conference is the Oil Minister of that country or another minister appointed in his place. If there is any uncertainty in the legality of a government during occupation there can be no 'Oil Minister' and therefore no representation of Iraq at OPEC. The criterion which OPEC themselves have applied is that only a minister from an interim government recognised by the UN can represent Iraq at an OPEC meeting.

Finally, Iraq is facing an enormous reconstruction programme, the energy requirements of which will be very high, possibly in excess of what the country can itself produce by way of crude oil. This might create a shortage of crude available for export, and by definition, an OPEC country has to be an exporter.

Iraq, Oil and Gas Union of

There has been a trade union presence in the Iraq oil and gas industry ever since the first production at **Kirkuk** in the 1930s. By 1973, the year the Iraqi oil industry was nationalised, the Union had nearly 50 000 members.

Reconstruction is hoped for in Iraq, measures including a new draft constitution. The Oil and Gas Union, using as its mouthpiece the Iraqi Federation of Workers' Trades Unions, is resisting any new regime under which the oil assets of Iraq could become privatised and pass into the hands of major foreign oil companies who, according to some reports, are eager to acquire oil fields in the region. Not only is there past and present western involvement in Iraqi oil—Shell and BP both currently hold exploration contracts—but foreign companies, notably Halliburton, Brown and Root, have been brought in to help repair the war-ravaged infrastructure of the industry. What the Oil and Gas Union of Iraq are fighting is possible *ownership* of the oilfields by foreign companies in a reconstituted Iraq.

Iraq, oil reserves in

Iraq has proven oil reserves of 112 billion **bbl** (9.9% of world reserves) and ranks second in the world behind Saudi Arabia. However, it is thought that up to 90% of the country remains unexplored due to years of wars and sanctions. Unexplored regions of Iraq could yield an additional 100 billion bbl. Iraq's oil production costs are among the lowest in the world. However, only about 2000 wells have been drilled (compared to over 1 million wells in Texas). Iraq is an isolated member of the **OPEC** group of countries. It has been excluded from quota allocations since 1990, prior to the first Gulf War. Despite the effects of the second Gulf War in early 2003, Iraq managed an annual daily oil production level of 2 million bbl/day. OPEC has set Iraq's production targets of 2.2 million bbl/day in 2007, 2.5 million bbl/day in 2008 and 2.8 million bbl/day in 2010. Historically, it had a production capacity of 6 million bbl/day and further investment could enable Iraq to ramp production up to 10–12 million bbl/day beyond 2012.

Irkutsk

(1) Russian oil tanker which on a Saturday in mid-2005 collided with a bridge on the Neva River in St Petersburg. A proportion of its payload of ≈ 3000 tonnes of diesel was released. The diesel leakage was less than a kilometre from the Hermitage Museum, one of the city's major weekend venues. The tanker was towed away after the collision.

(2) Site of the largest gas field in Russia, as yet untapped, known to contain $\approx 10^{12}$ m^3 of non-associated gas. There are various proposals for its development including a pipeline to China and Korea. At present, the local area uses an ageing coal-fired plant to generate power, so once production is underway at the Irkutsk field, 20% of its output will be allocated to local needs. However, much investment in infrastructure will be needed before any benefits can be realised. BP-Amoco have a significant interest in the field.

Islay

Island off the west coast of Scotland and one of the few scenes of power generation from **marine wave energy**: an **oscillating wave column** contributes to the grid. Most such generation worldwide is still only exploratory or experimental. Each turbine generates 500 kW. The commercial viability and funding for expansion have been secured by 15-year agreements from the power suppliers of Scotland to purchase the electricity generated at Islay.

Isooctane

Hydrocarbon (C_8H_{18}) of the alkane family having structural formula:

$$H_3C \overset{\displaystyle CH_3}{\underset{\displaystyle CH_3}{\vert}} CH_2 - \overset{\displaystyle CH_3}{\underset{\displaystyle CH_3}{CH}}$$

Its name according to the IUPAC convention is 2, 2, 4-trimethylpentane. It is used to determine the **octane rating** of gasolines.

Itaipu, Paraguay

Scene of major **hydroelectricity** in Paraguay, providing 94% of the country's 2.5×10^{12} Wh (≈ 10 TJ) annual demand and, perhaps more interestingly, one of the largest hydroelectric facilities in the world. Its capacity is 14 GW, some of which is exported to other Southern Cone countries. There are two smaller hydroelectric facilities in Paraguay, at Yacyreta and at Acaray. Paraguay has no known oil or natural gas reserves.

Italy, oil and gas reserves in

The first oil production in Italy was in 1861 and there was modest production up to the 1950s when a number of on- and offshore fields, the latter largely in the Adriatic, were discovered. Italy now produces about 5 Mt yr^{-1} of crude oil and **natural gas liquids**. Italy's natural gas reserves are 200 **bcm**. About 70% of the natural gas consumed in Italy has to be imported and provision for expansion of such importation is being made by construction of the terminal at **Campomento, Spain**.

Ivory Coast, oil production in

This politically unstable African country produces about 60 000 **bbl** per day of crude oil. Major onshore fields have included Espoir and Bélier, operated by Phillips Petroleum and Exxon respectively. Downstream activity is undertaken by the Ivorian Refining Company which has a refinery at Vridi. Up to the commencement of oil production in the Ivory Coast in the 1970s, the country had been dependent on coffee and cocoa exports, markets for which became very unstable at that time. Discovery and exploitation of the oil reserves were therefore timely. There is currently significant Indian investment in oil exploration in the Ivory Coast.

J

Jacket

Term originating at the Gulf of Mexico offshore industry and also used for North Sea installations, meaning a steel support for a fixed production platform. Such a structure is made from welded steel pipe and is attached to the seabed. Most oil and gas production platforms in UK waters, including three of the four at **Brent field**, use jacket support, the alternative to which is concrete support. A jacket typically weighs $\approx$20 000 tonnes and will support an approximately equivalent weight above the water surface. A jacket requires corrosion protection in the same way that sub-sea pipelines do. Common parlance in the offshore industry is 'topside' for the part of a platform above the water.

Jahre Viking

Japanese built, Norwegian owned **supertanker**. It was sunk by a missile during the Gulf War, salvaged and converted to an oil storage facility.

Jamaica, energy scene in

This country suffers from a paucity of hydrocarbon resources. There is an arrangement whereby it gets oil from Mexico and refined material from Venezuela on advantageous terms. There are proposals to install a **liquefied natural gas** terminal in Jamaica so that it can receive this product from Trinidad and Tobago, thereby reducing its dependence on imported oil. There is also some interest in **ethanol** as a fuel.

Japan Exploration Company (JAPEX)
Formed in 1955 with its HQ in Tokyo. At that time Japanese imports of crude oil were soaring and the refining capacity of the country enormously expanding. Onshore fields discovered at Yoshi and Katakai became productive in the 1960s and others followed. Two offshore fields came into operation in the 1970s and there are currently a total of seventeen fields in Japan or in Japanese waters producing oil, condensate and natural gas. The contribution that these make to the energy needs of Japan are close to being negligible: current figures are 5.5 million barrels per day of crude oil used by Japan of which a mere 5000 barrels are produced domestically, or just under 0.1 % of the requirement. Accordingly, JAPEX derives some of its profits from the provision of infrastructure for imported fuels e.g. pipelines for **liquefied natural gas** which Japan buys from sources including Alaska. JAPEX has not limited its activities to Japan but has had involvement in exploration and production projects in countries from which Japan purchases crude oil, including Canada and the United Arab Emirates. It is also represented in the **Sakhalin Energy** project.

Japan, importation and storage of autogas by
Japan is the world's largest user of **autogas (liquefied petroleum gas, LPG)** for automotive use and imports most of it from certain **OPEC** countries. Emergency storage facilities, capable of holding 650 000 tonnes of LPG, have been installed so that road transport of foodstuffs and other essential products will not be threatened if there is a contingency in the Middle East affecting LPG exports. It is estimated that, in the event of a cessation of imports, the quantity in storage would tide the country over for about two and a half weeks.

Japan's neighbour Korea is the world's second largest user of autogas, most of it imported from Dubai.

Japan, present and past consumption of motor fuel
In the early 21st century, Japan consumes over three hundred million barrels per year of gasoline. At the time that Japan started to prepare for World War II the motorists of Japan consumed a mere 6–7 million barrels per year: that was the allocation for 'civilian petrol consumption' immediately before the War. Some of that fuel would have come from the Dutch East Indies/Indonesia and would have gone into mo-

torcars manufactured by Nissan, as today. The company, previously Nihon Sangyo, was formed in 1934 and by 1937, when the 6–7 million barrels per year allocation of gasoline would have applied, was manufacturing 10 000 cars per year for the home market. As a result of the war, the allowance for civilian gasoline was cut to 1.6 million barrels per year, not all of which would have found its way into Nissan cars; Toyota had been producing cars since 1937, Mitsubishi since 1917. In the 1930s Mazda were producing only three-wheel motor vehicles for export to China, rather like the tuk-tuks which one still encounters in Bangkok, and did not make conventional cars until well after the War. Honda did not start making vehicles until the post-war years.

Jatropha tree
A source of **biodiesel**. Jatropha trees occur in African countries including Ghana and have been introduced in a number of Asian countries. Beans from the Jatropha tree produce oil at 40% yield which can be made into **biodiesel** by refining. The Jatropha tree can grow under poorer conditions of irrigation and nutriment than those required for the production of **rapeseed**.

Jefferson County, Alabama
The location of mining of **bituminous coal** which is particularly suited to **coke** production. Consequently Drummond Company Inc. who own the mine have 132 **coke ovens** with a combined output of $\approx$750 000 tonnes per annum and are the largest producer of coke in the US. Coal from this source is also used in power generation and some is gasified. Interestingly, the Warrior River passes through the company's land and barges are used for coal conveyance.

Jeruk field
Oil field in Indonesian waters shortly to be developed. Another Indonesian field soon to be developed is Cepu. The potential of these fields is important to Indonesia. The country's **OPEC** membership was in jeopardy in 2005 because she was not meeting the quota of oil production required for full membership. When the oil output from a country is assessed for the purposes of OPEC membership condensate is excluded.

Jesse, Nigeria
Scene of over 500 deaths in 1998 when a fire broke out at an oil pipeline. Local residents had been standing close to the pipeline with buck-

ets and bottles to scavenge oil from the pipeline. The fire spread to homes in nearby villages, killing families asleep in their homes.

Jet aircraft, fuel for

A jet engine is a gas turbine for which any fuel can be used, in principle, to provide carburetion at the point where it is required in the cycle. With jet aircraft, the fuel always used is petroleum material in the kerosene boiling range, although obviously much more precise specifications apply to jet fuel than to household kerosene. The flashpoint is an important property for a jet fuel: a high flashpoint is required so that in the event of fuel loss on landing there will not be ignition. If an aircraft is preparing for an emergency landing and it is carrying a large amount of fuel it might eject most of its fuel over the ocean before landing so that there will not be much left to leak if tank breakage occur during the landing.

Prices of jet fuel, like those of gasoline and diesel, fluctuate with the price of crude oil, which is why an **airline fuel surcharge** can legally be applied if between the booking of a ticket and the time of travel, crude oil prices take a surge. The **calorific value** of jet fuel is about the same as that of gasoline and diesel, ≈ 45 MJ kg^{-1}. Prices to airlines depend upon location, there being significant price variations between countries. Almost always the price also depends on the quantity supplied: bonuses of about 15–20% apply to the filling of almost empty tanks (equivalently, penalties apply when the tanks are merely topped up). As an example of the price of jet fuel, in July 2005 at Dallas/Fort Worth Airport, jet fuel is 89 cents per litre for a complete fill and $1.05 per litre for a small top-up. These prices contain some tax. Note the broad agreement with gasoline prices, which is to be expected in view of the common origin (crude oil) and comparable refining expenses.

Jevons paradox

An observation by William Stanley Jevons (1835–82) who stated that as technological improvements increase the efficiency with which a resource is used, the total consumption of that resource may increase rather than decrease.

Jevons observed in his 1865 book *The Coal Question* that England's consumption of coal soared after James Watt introduced his coal-fired steam engine, which greatly improved the efficiency of Thomas Newcomen's earlier design. Watt's innovations made coal a more cost-effective power source, leading to increased use of his steam engine in a wide range of industries. This in turn made total coal consumption rise, even as the amount of coal required for any particular application fell.

Jevons' Paradox is sometimes seen as a reason to not increase efficiency (if a particular resource is not used here, it will simply be used elsewhere). While this may be true, it does not take into account benefits the resource could generate for other customers e.g. a more efficient steam engine meant many more people could travel. Also, this principle is often referred to in conjunction with Peak oil, to show why conservation of oil will not slow the arrival or the effects of peak oil (see **Hubbert's Peak** theory which states that the rate of oil production on Earth will enter a terminal decline). However, a key part of Jevons' Paradox assumes a relatively steady supply of a given resource. Under this principle, demand increases after the price is reduced because of a fall in demand. Starting with a significant reduction in supply however (as in the case of Peak oil), prices will go up, requiring an equally significant reduction in demand from increased efficiency just to maintain the status quo of price and therefore consumption.

Jiyeh
Lebanese town south of Beirut and the location of an oil fired power station capable of producing 350 MW of electricity. During attacks on Lebanon on July 14th and 15th 2006, the power station at Jiyeh was bombed and 20 000–30 000 tonnes of fuel oil was released into the Mediterranean. The resulting slick threatened Turkey, Greece and Cyprus and effects on marine life were major. There was not only leakage of oil but also burning of some of it and destruction of the storage facilities. Soot and other debris therefore added to the environmental effects of the bombing.

Johor
Region of southern Malaysia located a few miles from Singapore, of importance to fuel and energy professionals because it not only receives many oil-bearing vessels but has become a repository for tankers which are out of service or awaiting a decision to scrap. Up to about 20 such vessels might be berthed in Johor at any one time. A spin-off is that in Johor, petroleum tanker sludge is being used as fuel in a **fluidised bed**, releasing heat at ≈ 12 MW.

Johor is in a Free Trade Zone, that is, a location where tariffs and quotas applying elsewhere in the country are waived, and has become an international trading centre for crude oil. **Qeshm Island**, Iran, is also a Free Trade Zone as is **Djibouti**.

Jordan, electricity supply in

The first electrical power in Jordan was in 1938. Current capacity is about 550 MW from thermal stations operating on a **Rankine cycle** using imported fuel oil, at present obtained from the **Daura Refinery, Baghdad**, to raise the steam. There is exchange of electrical power between Jordan and Egypt via a sub-sea cable in the Gulf of Aqaba and **Iraq** is currently seeking to import electricity from both countries. It seems probable that if this proposal comes to fruition, the majority of the exported power will be from Egypt by reason of its **hydroelectricity**: there are no hydroelectric installations in Jordan.

Joule

SI unit for energy or work. Its formal definition is the work done in moving a force of 1 Newton a distance of 1 metre.

Julich, Germany

Scene of a small (13 MW of electricity) nuclear power station which operated only intermittently over its twenty-year life, usually with **thorium** as fuel. **Burnup**s of up to 150 GW day per **tonne of Heavy Metal (tHM)** were achieved.

Jurong Island

Scene of an industrial estate in Singapore on land which was previously marsh. **Singapore Petroleum Company** have a refinery there with a throughput of 309 000 **bbl** per day of crude oil. Shell and BASF each have a presence on Jurong Island.

K

Kalimantan
The part of the island of Borneo which belongs to Indonesia and the scene of rapidly developing coal production in particular at the Kaltim Prima mine. Coal from this mine is bituminous with a **calorific value** as received of 28.5 MJ kg⁻¹ It is low in mineral matter and in sulphur. Production began in 1992 at a level of 7.3 Mt yr⁻¹, recent production being just under three times that, most of it for the export market.

Kamchatka
Scene of **geothermal electricity** generation in Russia, there being only a few small installations for such generation in a country of significant geothermal resources. It is doubtful whether the total power from geothermal generation in Russia exceeds 100 MW.

Kansas, ethanol production in
Largely because of its abundance of corn, Kansas is becoming a major player in the production of **ethanol** for fuel use. **Sorghum** is also used in the in the State's seven ethanol plants. The process involves production of carbohydrate and fermentation, and corn and Sorghum can be used interchangeably. Production was 170 million gallons in 2005, some of which was sold in the form of **E85**.

Kapok

An insulating material used in combustion plant, for example, between the inner and outer metal walls of a furnace. Though an effective insulator, it has the disadvantage of being combustible and therefore a contributor to the fire load of the area in which it is in use. Its chemical composition is similar to that of cotton.

Kapuni

Onshore gas field in New Zealand distinguished from most others by the low proportion of methane in the gas—a little under 50%—giving it a calorific value of only about 25 MJ m^{-3}. This disadvantage is to some extent offset by relatively large amounts of C_{3+} hydrocarbons, about 3%. This is stripped off as **natural gas liquids.**

Kariba Dam

The location of major hydroelectric activity in Africa. Power generation at the facility began in 1960, and the border between Zambia and Zimbabwe passes through it. The south side is in Zimbabwe and the north in Zambia; each has a generating capacity of 600 MW. Some of the power generated is exported to neighbouring countries.

Kela-2

Non-associated gas field in northwest China. Gas production there began in 2004 with **Sinopec** as operator.

Kenai, Alaska

Location of the first **liquefied natural gas (LNG)** terminal in the US, which began operations in 1969 and is still functioning. In fact it is still the only LNG export terminal in the US: there are four import terminals and about 100 storage facilities, some of them receiving natural gas and liquefying it on site. The destination of all of the despatches from the Kenai terminal is Japan.

Kenting National Park, Taiwan

The southerly tip of this island nation, it attracts many tourists but is also the site of a nuclear power plant. Both the Park's wildlife and the operation of the nuclear plant were threatened when in 2001 a Greek registered vessel in nearby waters released 1150 tonnes of its fuel. There was also the possibility that if the leaked fuel drifted west, reprisals from China would follow.

The nuclear power plant in the Park was the third to be built in

Taiwan. The two earlier plants are at the northernmost part of the island.

Kentucky, natural gas in

This landlocked state of the US has long been a supplier of natural gas. Gas from some deposits in Kentucky tends to be high in ethane, as might be expected from the fact that there is also a layer of shale at such gas deposits. Three further points of interest might be made. First, the gigantic reserve of shale known to pass from northeast to southwest in the US landmass passes through Kentucky. Secondly, the only natural gas ever utilised to have a predominance of ethane over methane was from Kentucky (as far as the authors are aware, such gas is no longer produced). Thirdly, raw shale oil made in Kentucky by retorting has been adapted by **steam reforming** to **town gas**. To do this with crude oil would be wasteful, but raw shale oil can be so gasified economically.

Kenya, oil and gas situation in

When Imperialism was finally surrendered in the African continent in the post-war years, Kenya was one of the few newly independent nations to become relatively prosperous while its neighbour Tanzania, for example, continued as a subsistence-level farming economy. A barrier to continued growth in Kenya has been the fact that the country has no known oil and gas reserves. Accordingly, the country has granted on- and offshore exploration licences to companies including Global Petroleum, Woodside Petroleum and Dana Petroleum. Activity to date has revealed no provable reserves but has provided a basis for cautious hope that there will be worthwhile amounts of oil off the Kenyan coast. Exploration wells are on the immediate agenda.

Imports are currently meeting the country's oil needs, which are a modest 35 000 **bbl** per day representing static circumstances. In order to have energy for growth and development, Kenya will need domestic oil.

KeySpan Energy

Distributor of natural gas for northeast US, also a producer of electricity for New York City and Long Island. Electricity so produced is sold to power companies for distribution. Some of the natural gas distributed by KeySpan is imported from Canada. There was a break-in at a KeySpan **liquefied natural gas** storage facility in Massachusetts in August 2006, which went unnoticed for several days.

Khodorovsky, Mikhail

Former Chief Executive of the **YUKOS Oil Company**, in January 2007 sentenced to eight years in a Siberian Prison on tax evasion charges. He was succeeded by Simon Kukes, a US citizen of Russian birth who had previously held posts with companies including ConocoPhillips. His remit will be to restore investor confidence in Yukos after the recent difficulties.

Kieselguhr

A mineral substance (a form of silica composed of the shells of diatoms) which can be used in the manufacture of **refractory materials**. When mixed with fireclay, kieselguhr has the effect of making it less dense and more porous. Fireclay so modified is seldom used as the inside surface of a **firebox**. It will more commonly be used as an insulator between the **firebox** and the outside surfaces of the combustion plant.

Kikeh, Malaysia

Scene of exploration for **deep-sea oil and gas** in eastern Malaysian waters. The depth at which drilling will occur is 1330 m and estimated reserves are up to 700 million **bbl** making Kikeh technically a **giant field**.

Kikeh Field

Malaysian oil field with a potential yield of 120 000 **bbl** per day. Expected to begin production in 2007, it will be operated by the US concern Murphy Oil.

Kiln

A kiln is a thermally insulated chamber in which a controlled temperature regime is produced. The design of the chamber normally focuses on insulation, and the ability to add fuel over a period of time. Care must be taken not to heat the kiln too rapidly or to too high a temperature. Kilns are used to harden, burn or dry materials. Specific uses include:

- converting wood into charcoal or to dry green lumber so that the lumber can be used immediately
- annealing, fusing and deforming glass
- cremation (at high temperature)
- drying of tobacco leaves
- firing of certain materials to form ceramic materials.

Kilve, UK

The location of an unsuccessful fuels enterprise in Somerset in the 1920s. A Dr Forbes-Leslie claimed to have identified a shale deposit, which in itself was altogether reasonable as by that time shale had been excavated and made into liquid fuels in several parts of Britain. Exaggerated claims of the extent and quality of the shale in Somerset were however made and in spite of the installation of retorting, refining and rail transport facilities, the enterprise came to nothing. The sole memorial to it now is the retort which has existed with redundant status for about 80 years.

Forbes-Leslie's initial promotion had encouraged the uninformed to believe that something comparable to Spindletop had occurred in the English rural county. Had his claims been realised, not only would the oil have been an asset to the Navy in particular, but many jobs would have been created during a time of high unemployment. It should be noted that one of Forbes-Leslie's subsequent enterprises earned him a period in gaol.

Kinder Morgan

Houston based pipeline operator currently responsible for 35 000 miles of pipeline. The company is proposing a pipeline which if it comes to fruition will significantly exceed in length even the **Baku-Tbilisi-Ceyhan Pipeline**. It will convey natural gas from the Wyoming fields to manufacturing plants in eastern Ohio. Part of the rationale is that sales of the Wyoming gas in California will decrease as new **liquefied natural gas** terminals, including the terminal in Baja California, come into operation therefore new markets for the Wyoming gas must be sought.

King Cove, Alaska

Scene of hydroelectricity generation at a very small scale: 800 kW. The installation, which has been in operation since 1994, serves a remote community of fewer than 1000 persons.

Kingfish

Largest of the oil fields in the Bass Strait off south eastern Australia. A **wildcat** when first drilled, it has yielded over a billion barrels of oil since it began production in 1971.

Kinston, North Carolina

The scene of a fatal dust explosion at a pharmaceutical plant in 2003. Substances such as sugar, grain and cereals which we do not ordinarily consider to be fuels are capable of dust explosion if dispersed in air within the flammability limits.

Kirkuk

Location of the first discovery of crude oil in Iraq in 1927. Production did not begin until several years later, by which time Bahrain had become the first oil producer in the region.

Kiushui

Site of the first significant coal production in Japan early in the industrialisation of the country. Much environmental harm resulted, including water pollution and damage to crops. More recently Kiushui has been a centre of **geothermal electricity** generation. In fact all of Japan's geothermal electricity is produced either at Kiushui or at Honshu. The total capacity is about 570 MW although it is known that the geothermal resources of Japan are intrinsically capable of raising power at 2.5 GW.

Knock

Knocking, technically detonation, in internal combustion engines occurs when the fuel/air mixture in the cylinder has been ignited by the spark plug and the smooth burning is interrupted by the unburned mixture in the combustion chamber exploding before the flame front can reach it. Combustion stops suddenly, because of the explosion, before the optimum moment of the four-stroke cycle. The resulting shockwave reverberates in the combustion chamber and pressures increase catastrophically, creating a characteristic metallic pinging sound. If allowed to persist, detonation will damage or destroy engine parts.

Detonation can be prevented by:
- the use of a fuel with higher **octane rating**
- the addition of octane-increasing lead, **isooctane**, or other fuel additives
- reduction of cylinder pressure by increasing the engine revolutions, decreasing the manifold pressure or reducing the load on the engine, or any combination
- reduction of charge temperatures (such as through cooling, water injection or compression ratio reduction)
- retardation of spark plug ignition
- improved combustion chamber design that concentrates the mixture near the spark plug and generates high turbulence to promote fast even burning

• use of a spark plug of colder heat range in cases where the spark plug insulator has become a source of pre-ignition leading to detonation.

Correct ignition timing is essential for optimum engine performance and fuel efficiency. Modern automotive and small-boat engines have sensors that can detect knock and delay the spark plug firing to prevent it, allowing engines to safely use petrol of below-design octane rating, with the consequence of reduced maximum power output and efficiency.

Known recovery

A term for the total past production of an oil or gas field and current estimate of proven reserves.

Knutsen/NaturgasVest

Developers and operators of the most advanced vessel for **liquefied natural gas** transportation currently in existence. Small in size, it currently only operates on waterways within Norway. However, some of its features will undoubtedly be applied to larger vessels of the future. The most important is the use of gas which has boiled off from the payload as fuel to power the vessel itself. Previously such boil-off had been seen as a loss, the vessel being powered with fuel oil.

Koho Maru 5, (also known as Pak-1)

A vessel for **liquefied petroleum gas** transport, manufactured and commissioned in Japan in the early 1970s. It was used for its intended purpose until it purchased by a Thai company, re-registered to Thailand and renamed Pak-1. The vessel sank in the Gulf of Thailand in the mid-1990s after a collision with a Thai Navy vessel. Having started to drift, causing a collision hazard, the wreck of Pak-1 was eventually towed to a part of the Gulf away from shipping lanes, and blown up with high explosives.

Koolmotor

Trade name of one of the many gasoline-alcohol blends for spark-ignition engines that have been available in Britain over the decades. Others include Benzalcool, Moltaco, Lattybentyl, Natelite, Alcool and Agrol.

Korean Coal Corporation (KOCOAL)

Like Japan, Korea has significant coal reserves. The Korean Coal

Korean National Oil Corporation (KNOC)

Corporation (KOCOAL) was formed in 1950, the year the Korean War began, and with its HQ in Seoul has operated three mines on behalf of the government, which annually produce about 1.2 Mt of anthracite. The Korea Electric Power Association purchases anthracite from KOCOAL.

Korean National Oil Corporation (KNOC)
South Korea is the world's fourth largest importer of crude and KNOC is concerned with investment in overseas energy projects, affording the country a place and a voice in the oil and gas industry. For example, KNOC has entered into an agreement with Iraq whereby South Korea will participate in exploration and development projects in the Iraqi oilfields. KNOC has also obtained Korean involvement for projects in Yemen, Indonesia, Vietnam and Peru.

Kudu gas field
An offshore non-associated gas field in Namibian waters with proven reserves of 35 **bcm** of natural gas. At present, development is under way whereby gas from the field will be piped to Oranjemund on the mainland and used to generate power at 800 MW. Some is exported to other countries in the region.

Kurdish people, assertion of rights to oil revenues in Iraq
Because of historical events, the Kurds became a race dispossessed of territory. Kurdistan is the region in which they live but it has no political or constitutional basis. Kurds are Shiite Muslims with their own language and distinctive culture, and currently number 15–20 million. Many Kurds have returned to **Kirkuk** and are asserting oil rights but Turkey has also stated its interest and rights in the oil resources of Kirkuk.

Kuwait, loss of crude oil during the Gulf War
During the 1991 Gulf War, oil fields in Kuwait were torched by Iraqi engineers by order of Saddam Hussein, with the loss of a billion **bbl** of crude oil and almost inestimable health and environmental harm. The burning rate was up to 4 million bbl per day and smoke drifted as far as Turkey, Syria and Afghanistan. A further 10 million barrels of oil were released into the sea. By the beginning of the 21st Century, there had been little by way of clean-up and there were many incidences among Kuwaitis of cancer and other illnesses attributable to the atmospheric contamination.

Several refineries in Kuwait including that at **Mina al-ahmadi** suffered damage as a result of the Gulf War.

Kwazulu-Natal coast, hydrocarbon activity at

The first refinery in South Africa was sited in the coastal town of Durban, commencing production in 1954. This refinery is still in operation, processing 125 000 **bbl** per day. It is now owned by Engen, who are a subsidiary of the Malaysian national petroleum company Petronas. About ten miles out of Durban there is a refinery operated by South Africa Petroleum Refining (SAPREF) of which both Shell and BP have a significant share. This processes 180 000 bbl of crude per day. There is an oil storage facility in Durban harbour owned by SAPREF. In November 2002 a leak of about 95 bbl of oil from this facility into the sea took place. There have been other cases of leakage of liquid hydrocarbon into Durban harbour over the last few years.

Kyoto Protocol

An international agreement setting limits on emissions of carbon dioxide and other greenhouse gases. It originated at an assembly in Kyoto, Japan, in 1997. Targets vary by region: in the UK a 12.5% reduction on the 1990 emissions by 2012 is aimed for. What were initially recommendations were actually enforced in 2005 in the countries—141 in all—which had ratified the Kyoto protocol. The US had not ratified it, neither had Australia which nevertheless has its own quite ambitious CO_2 reduction targets. Countries which have ratified the agreement but where the industrial milieu is either emergent or underdeveloped are excluded from target CO_2 emission reductions for the time being; Brazil is a case in point.

La Brea

Reserve of **bitumen** in Trinidad. About 15 000 tonnes are recovered annually, most of it for export.

La Muela wind farm

Near Zaragoza in Spain, and the first wind farm in Europe to have been operated by Shell. It has an output approaching 100 MW.

La Preciosa coal mine

In northern Colombia, the scene in February 2007 of an explosion in which 32 miners were killed. The story was a familiar one: methane leakage and an ignition source in the form of a spark. It was at least the third such accident in Colombian coal mines in a decade.

Lahti, Finland

Scene of a **fluidised bed** reactor which pyrolyses wastes of various sorts including **municipal solid waste**, plastic waste and **biomass**. The gas is passed along to the plant which raises steam for power and has burners both for natural gas and for **pulverised fuel (pf)**. Gas from the reactor is co-fired with natural gas and the pf combustion cut back by a thermally equivalent amount. Coal consumption at the power plant is thereby reduced by 45 000 tonnes per annum.

Landfill gas (LFG)

Gas composed of 45–65% methane, balance inerts, created by the decomposition of rubbish at landfills on a time scale of 15–25 years. There have been applications to power generation using a **genset** especially developed for the purpose which has a spark-ignition engine and is capable of delivering power at up to about a megawatt. One site in the northwest of England, which uses 18 such gensets powered by LFG, is amongst the largest such facility in Europe. Landfill gas not utilised can be burnt at a **flare**. This is environmentally advantageous, as carbon dioxide is a less powerful greenhouse gas than methane. There has also been significant use of LFG for power in Hawaii, using a gas turbine.

Laos, electricity production in

There are several scenes of **hydroelectricity** in this Asian country with a capacity of 20 GW. Most of it is sold to Thailand, but in that region of the world there are plenty of alternative purchasers if the demand for exported power to Thailand declines because of **Mae Moh**, for example. It is expected that within a few years a common grid will provide power to at least some of the Association of South East Asian Nations (ASEAN) countries and Thailand is in fact planning to raise its generating capacity to that end. Thailand was one of the founding members of ASEAN in 1967: Laos joined in 1997. ASEAN countries collectively stand to benefit from the 'common grid' proposals.

Latrobe Valley

Place in the Gippsland region of southeast Australia and the scene of much **brown coal** production from several open cuts. The coal is used locally to make electricity for the State of Victoria and there is also a major plant for producing **briquettes**. There is also activity in commercial **char** production and in the conversion of coal to liquid fuels.

de Laval nozzle

Convergent-divergent nozzle used in **rocket** propulsion. Its inventor Gustav de Laval founded the company AB Separator in 1883, later renamed Alfa Laval and now a huge multi-national concern.

Lavera Terminal

On the French coast near Marseille, having a storage capacity of 180 000 m^3 of liquid hydrocarbons. There are a number of refineries close to the terminal with a combined capacity of 600 000 **bbl** per day.

Some crude oil received at the terminal, often via the Suez Canal, is exported onwards, while some is refined locally and returned to the terminal for export to the Mediterranean region and west Africa.

Law of Capture

Applying in the early days of the US oil industry, this law stated that the oil which any producer brought to the surface of a well on his own land (which could mean owned or leased) was his. Clearly, such oil might well have been beyond the boundaries of the producer's land before being drawn to the well, so producers tended to site wells on their land so as to obtain crude from as large an area as possible. The result was that sometimes an entire reservoir would experience an internal pressure drop because of the activities of one producer on a relatively small piece of land, and this made the remaining oil in the reservoir more difficult to bring to the surface and so reduced the value of the reservoir and jeopardised the prospects of subsequent drillers with land above the reservoir.

Although the Law of Capture as such was not revoked, new laws came into being in the 1930s whereby regulatory authorities had the entitlement to check the oil-to-gas production ratio of any well. A low value of this ratio signified a reduction of reservoir pressure. A well with a lower value of the oil-to-gas production ratio than that set by State law would be deemed to be a threat to the reservoir and compulsorily closed down.

Lead, tetraethyl

Once a very widely used octane enhancer, now largely proscribed because of the environmental effects of lead from car exhausts. Engine knock is caused by ignition ahead of the spark, possibly involving **cool flame** behaviour, and the effectiveness of tetraethyl lead was due to its ready release of ethyl radicals which would scavenge reactive intermediates which otherwise would have built up to sufficient proportions to cause knock.

Lead-acid battery

Lead-acid batteries, invented in 1859 by French physicist Gaston Planté, are a type of galvanic cell and are the most commonly used rechargeable batteries today. They also represent the oldest design with one of the lowest energy-to-weight ratios, commonly around 30 Wh/kg. The energy-to-volume ratio is also low compared to other types of batteries. The power-to-weight ratio can be quite high, however. They are relatively low-cost and can supply high surge currents needed in starter motors. Every reasonably modern car uses a lead-acid battery for this purpose. They are also used in vehicles such as forklifts, in which the low energy-to-weight ratio may in fact be considered a benefit since the battery can be used as a counterweight.

Lead-acid car batteries for a 12 volt system consist of six cells of 2.1 V nominal voltage. Each cell contains, in the charged state, electrodes of lead metal (Pb) and lead (IV) oxide ($Pb^{IV}O_2$) in an electrolyte of about 37% (or 6–12 M) w/w sulphuric acid (H_2SO_4). In the discharged state, both electrodes turn into lead (II) sulphate ($Pb^{II}SO_4$) and the electrolyte turns into water.

The chemical reactions are, charged to discharged:

Anode (oxidation): $Pb_{(s)} + SO_4{}^{2-}{}_{(aq)} \leftrightarrow PbSO_{4(s)} + 2e^-$

Cathode (reduction):

$PbO_{2(s)} + SO_4{}^{2-}{}_{(aq)} + 4H^+ + 2e^- \leftrightarrow PbSO_{4(s)} + 2H_2O_{(l)}$

Because of the open cells with liquid electrolyte in most lead-acid batteries, overcharging with excessive voltages will generate oxygen and hydrogen gas by electrolysis of water, forming an extremely explosive mix. Caution must also be observed in handling because of the extremely corrosive nature of sulphuric acid.

Lead-acid batteries are used in emergency lighting in case of power failure since they react less violently to fire exposure than nickel-cadmium batteries.

Attempts are being made to develop alternatives to the lead-acid battery, particularly for automotive use, because of concerns about the environmental consequences of improper disposal of old batteries. Lead-acid battery recycling is one of the most successful **recycling** programs in the world, with over 97% of all battery lead recycled between 1997 and 2001. An effective lead pollution control system is a necessity for a sustainable environment. There is a continuous improvement in battery recycling plants and furnace designs for greater efficiencies. These recycling plants are ecologically friendly as they follow all emission standards for lead smelters, and new methods are being developed so that lead pollution can be reduced to an essentially negligible amount.

Leather waste, gasification of
Leather waste in raw form can be gasified conventionally with air to make **producer gas**. Alternatively, the waste can be made into **briquettes**

which are gasified by pyrolysis leaving a usable **char** as the solid residue.

Lebanon, energy scene in

This country imports its entire consumption of a million **bbl** per day from countries including **Saudi Arabia** (by sea) and **Iraq** (by pipeline). Exploration for on- and offshore oil is in progress and the British Licensee Spectrum have discovered several oil reservoirs within 12 miles of the coastline which, it is hoped, will eventually eliminate the country's total dependence on foreign oil. Lebanon is changing from oil to natural gas for power generation although the fuel will have to be imported, by pipeline from Syria or from more distant sources as **liquefied natural gas**. Of course, this could change if the offshore oil wells currently being explored eventually yield some **associated gas**. Power generation in Lebanon is about 2000 MW produced by seven thermal power stations. There is also some import of electricity from Syria.

Leduc

Name of the first major oilfield of Canada, where production began in 1946. It is in Alberta, whereas previously utilised oilfields in Canada, having nowhere near the potential of the Leduc field, had been in Ontario. Development of the Leduc field stimulated growth of the Canadian oil industry, there being further major discoveries, most of them in Alberta and some in Saskatchewan. The country has been producing in excess of 1 million **bbl** per day of crude oil for nearly 40 years.

Lelystad, Holland

Scene of **combined heat and power (CHP)** using entirely wood from a **short rotation coppice** as well as from other sources including **forest thinnings**. The short rotation coppice is expanding alongside the CHP facility itself, there having been plantings of poplar and willow occupying about 200 hectares. Promoters of this arrangement have made the point that a short rotation coppice does not in any way make for an unattractive landscape. The tree plantation does not become disfigured when wood is removed from selected individual trees. Also, at any one time many trees in such a plantation are displaying powerful re-growth at the position of previous wood removal.

Leonardite

Term sometimes used, incorrectly, as a synonym for **lignite**. Like lignite it is formed, on a geological time scale, from **peat.** The difference is that lignite is devoid of micro-organisms and proceeds along the **coalification** sequence geochemically whereas leonardite has retained micro-organisms which have retarded its development into coal and given it a high content of humus. Hence its primary application is as a soil-enriching agent. It is also sometimes used as an ingredient of **drilling fluids**. New Mexico, USA has large reserves of leonardite.

Liberty Gap Wind Force
Project underway in West Virginia whereby fifty **wind turbines** will come into operation on a 6 to 7 mile stretch of mountainous terrain. The capacity will be 125 MW.

> Other wind energy projects in West Virginia include one in Greenbrier County where 124 turbines having a combined capacity of 186 MW are planned.

Liddell Power Station
Major thermal power station in New South Wales, Australia, where there is co-firing of sawdust with local **bituminous coal**. Australia produces 50 Mt yr^{-1} of **biomass** waste and co-firing is preferable to simple incineration or taking to a landfill.

Lignite
An approximate synonym for **brown coal**. Its origin is that some of the brown coal reserves of the US, unlike those in Germany where brown coal utilisation first began, have a fibrous texture traceable to the lignin in the initial plant deposit hence the word lignite. Some texts have attempted to impose a distinction between a lignite and a brown coal on the basis of moisture content as mined. It is usually safe to equate lignite to brown coal.

Limestone
Rock composed mainly of calcium carbonate, $CaCO_3$. It is used to remove sulphur dioxide from post-combustion gases by the process:

$$CaCO_3 + SO_2 + \tfrac{1}{2}O_2 + 2H_2O \rightarrow CaSO_4.2H_2O + CO_2$$

It is usual to use about 30% excess lime (that is, 30% more than required according to the stoichiometric equation above) in which case SO_2 removal efficiencies of 80% or better are possible. An example of application of this classical technology in 2005 is the burning of coal waste at **Bakerton, PA**.

Linz-Donawitz gas

Gas occurring as a by-product in steel manufacture having a **calorific value** of 9 MJ m^{-3}. It is put to use at the steelworks as a fuel gas, possibly as a blend with a richer gas such as **coke** oven gas.

Liptinite

A **maceral** group derived from the waxy and resinous parts of plants such as spores, cuticles, and resins, which are resistant to weathering and diagenesis. Very sensitive to advanced coalification, the liptinite macerals begin to disappear in coals of medium-volatile rank and are absent in coals of low-volatile rank. When the liptinite macerals are present in a coal, they tend to retain their original plant form and thus they are usually plant fossils or phyterals. The phyteral nature of the liptinite macerals is the main basis on which they are classified.

Liquefaction

The process of turning a solid or gas into a liquid. Research into the liquefaction of coal has been carried out for over a century. Rising oil prices and decreasing worldwide oil reserves in the not too distant future has renewed interest in coal as a starting material for liquid fuels and raw materials for the chemical industry. In the Bergius process for direct coal liquefaction, developed at the Max Planck Institute for Coal Research in Germany, the coal is treated with **hydrogen** under pressure (25 MPa, at 350°C) in a mixture of sodium borohydride and iodine in the solvent, pyridine. In this reaction, the carbon-carbon bonds between aromatic and aliphatic parts of the molecules are broken and the free bonds are saturated with hydrogen; the network structure of the coal is disrupted. In addition, the double bonds of the aromatic ring systems are partially hydrated so that the aliphatic content rises at the cost of the aromatic. By this process, the first true coal hydration was carried out in the sense of hydrogen being added to unsaturated structures and high-rank coals were liquefied in a conventional hydrocracking process for the first time.

Liquefied natural gas (LNG)

Natural gas having been converted to liquid form for the purpose of storage and/or transportation. The critical temperature of methane is such that (unlike propane) it cannot be made into a liquid by application of pressure at room temperature. The manufacture of LNG therefore requires that the gas be cooled below its critical temperature of 190 K. This is often done by passing the natural gas through a nozzle from which it emerges at high speed, enthalpy having been converted to kinetic energy and the necessary cooling having been attained. Liquefaction at atmospheric pressure occurs at 112 K and this is achieved by the use of successively colder refrigerants. LNG is therefore a liquid in equilibrium, or at least in contact, with its vapour at its boiling point. Natural gas is not of course pure methane and the composition in terms of inerts and the higher hydrocarbons such as **ethane** changes in the liquefaction process, although some of the constituents other than methane are retained. This can in fact lead to difficulties if the heavier components of the LNG descend to the base of the vessel and density gradients develop. This can lead to internal circulation and mechanical instability; there are well-documented cases of such occurrences. LNG obviously has to be converted back to gas for reticulation and supply and this is often done by heat exchange with seawater.

The initial *raison d'etre* of LNG was that suppliers of gas could buy it in liquefied form during the warmer months when it was cheaper and store it for reticulation to customers during winter. Inevitable evaporative losses of the gas during storage did not nullify the savings made from having bought the gas at the most advantageous time of year. This all came to an end in 1944 when there was a severe accident in Cleveland, Ohio due to failure at an LNG facility. The material evaporated and drifted into the streets: 128 persons died from the resulting fire and explosion.

A few years later LNG manufacture resumed and by now it is one of the world's staple fuels. It can be transported by land or, in specially constructed vessels, by sea. The largest producer of LNG is the US and the second largest Trinidad and Tobago. The world's largest exporter of LNG is Indonesia and there are other **OPEC** countries including Qatar which export LNG. There is also production on a large scale in the former USSR. Current annual world production of LNG stands at about 80 Mt. The world's largest importer of LNG is Japan. About 25% of world exports of natural gas are as LNG rather than by pipeline in gaseous form.

One of the attractions of LNG is that it enables natural gas from offshore fields to be collected and prepared for marketing even if there is no infrastructure for distributing the gas once ashore: this applies very much in Indonesia. There are alternative technologies for converting natural gas to a product which can be made into a liquid and these include **oxidative coupling** to make longer chain organics which are liquid at room temperature. They also include direct conversion of the gas to methanol by partial oxidation. In spite of international research and development with some initially promising results, there has been no commencement on a large scale. What *is* on the increase is the purchase of LNG as a feedstock for making methanol instead of for fuel use, for example at **Freeport, TX**.

Liquefied natural gas, transportation by sea

From the voyage of the **Methane Pioneer** in 1959 to the present day there is no record of a fatality or a serious injury in the transportation of **liquefied natural gas (LNG)** by sea. The distance covered by the LNG-bearing vessels of the world over that period is estimated as 60 million miles.

Liquefied natural gas, vehicular use of

Compression ignition engines which run on methane are well established, there being many makes and models of car designed to run on compressed natural gas. This of course is simply methane gas a long way above its critical temperature stored under pressure in the vehicle for supply to the engine. As an alternative, the gas can be carried as **liquefied natural gas (LNG)** and evaporated on its way to the engine. The fuel tank in such a vehicle comprises two membranes with an evacuated space between them, a scaled down form of the membrane insulation devices which are being used in the current generation of ocean tankers for LNG.

Whenever a new fuel for motor transport is introduced, the confidence not only of the consumer but also of the forecourt operators, whose collective influence is considerable, has to be won. The authors' view is that both LPG and **methanol** as fuels for vehicles were at least initially adversely affected by lack of such confidence. Consequently much research and development went into making refuelling facilities for vehicles powered by LNG not only user-friendly but familiar so that filling with LNG is, to the motorist, no different from filling with petrol.

Liquefied petroleum gas (LPG)

Gas composed of propane and **butane**, previously dissolved in crude oil, which comes over at the first stage of refining. The term LPG can in fact mean one of four things:

- the mixture of propane and butane as it naturally occurs
- the natural mixture adjusted to give a required composition e.g. 60% propane and 40% butane
- such a mixture with all of the butane removed therefore consisting of pure propane
- such a mixture with all of the propane removed therefore consisting of pure butane.

Unlike natural gas, LPG in any of the above forms can be converted to liquid by application of pressure only at ordinary temperatures and stored and transported in vessels, usually made of carbon steel, capable of holding the pressure of the vapour with which the hydrocarbon so liquefied is in equilibrium. Such pressures are in excess of 10 bar. Vessels for ocean transportation of LPG are of course purpose built and large oil companies including Shell tend to charter rather than own such vessels. Land transportation is most commonly by rail, usually from an LPG source or terminal to where it is required. Only in Canada and Russia is there long-distance carriage of LPG by rail.

The world's current largest producer of LPG is the USA and the world's largest exporter Saudi Arabia. Several other **OPEC** nations including the UAE and Indonesia are large producers and exporters, as are the UK and Australia. The world's largest importer of LPG is Japan with China and Korea closely following, countries with a high energy demand. In many countries including the UK, the US and Australia LPG finds significant use as a transportation fuel on a scale ranging from large freight vehicles and trans-continental coaches through family cars to forklift trucks. Propane/butane blends are favoured for such applications. LPG is also used in homes and in restaurants and hotels for cooking when propane alone, in the absence of butane, is preferred. LPG is used in the metallurgical and glass industries where in many applications it replaced **producer gas**. It is also used as a feedstock for petrochemicals as, in some places including the southeast US, is **liquefied natural gas (LNG)**.

Combustion phenomenology if there is accidental loss of LPG depends on the rate of leakage. If an orifice is created in an LPG vessel and the hydrocarbon so released ignites, a jet fire will result. If an LPG vessel breaks open so that its contents leak rapidly there will be a **boiling liquid expanding vapour explosion (BLEVE)**, followed, if there is ignition, by a fireball.

Liquefied petroleum gas (LPG), non-interchangeability of different types

Large amounts of LPG (pure propane) are used in cooking, and a blend of propane and butane for vehicular use (sometimes called **autogas**). These should not be interchanged and to attempt to do so can be dangerous. Autogas when burnt in air has a higher flame speed than pure propane, so if autogas is used on a cook-top burner adjusted for propane there can be flashback, i.e., propagation of the flame down the burner pipe work. Also, if autogas is used on a propane burner it will not entrain the correct amount of air. Possible effects of this include **yellow tip**.

Liquefied petroleum gas (LPG), tankers for ocean transportation

The simplest are called pressurised **LPG** carriers and carry tanks of LPG having the same design features as those used for storage on land. A step up is the semi-refrigerated type which subjects cargo having boiled off to a refrigeration cycle, preventing inventory loss and keeping the bulk LPG cool but still well above its boiling point. There are also fully refrigerated types in which the LPG is carried at about – 48°C where the cargo hydrocarbon itself is the refrigerant. The boiling point of propane is –42°C, therefore its pressure at –48°C is less than 1 bar so there is no need for it to be stored in a pressure vessel and the walls of the LPG-containing vessel can be thinner than those in either of the other types.

The hydrocarbon tanker fleets of the world are at the present time moving to **double-hulled tanker** configuration. At the present time there are still many single-hulled LPG vessels in service.

Liquid fuels from synthesis gas, economics of
At the **Sasol** plant near Johannesburg, liquid fuels are made from coal via **synthesis gas** as South Africa lacks oil reserves, so the manufacture there of liquid fuels from coal has a sound economic basis. In countries having domestic oil, there is a critical price of crude oil above which it becomes economic to manufacture liquid fuels from **synthesis gas**. Currently, this cut-off figure in relation to the US is said to be about $30 per **bbl** of crude oil while the actual price is twice that or more. Accordingly Montana, one of the northwest states bordering Canada, is assessing its coal reserves with a view to long-term production of liquid fuels which will compete with those from domestic oil. This will of course be a grass roots activity requiring major investment for initiation.

Lithotypes of coal
The four coal lithotypes are:
 • Vitrain, displaying glassy bright layers showing orthogonal cracks and angular grains
 • Clarain, less bright and glassy than vitrain
 • Durain, dull and hard layers with a granular texture
 • Fusain, characterised by fibrous layers.

Live biomass, release of carbon dioxide by
A frequently asked question in the matter of removal of carbon dioxide from the atmosphere by trees is whether they also release it be-

cause of their respiratory functions. The answer is yes they do, but much more slowly than they remove it by photosynthesis, so there is net removal. A plantation of trees can thus be seen as a **carbon seques-tration** facility.

Liverpool Bay oil and gas fields
These fields are only a few tens of miles from the **Morecambe Bay gas field**, occasionally referred to as South Morecambe, and are considered to be a 'North Atlantic' field. The operator is Broken Hill Proprietary (BHP), an Australian concern, and the Liverpool Bay development is currently their largest single activity. Oil and gas are brought ashore at the nearby Welsh coast. The life expectancy of the resource is estimated as 20 years, eventual oil production being 70 000 **bbl** per day, all of which will be exported.

Lobito
Scene of a proposed **grass roots refinery** for Angola with a projected capacity of 200 000 **bbl** per day. Lobito is a coastal locality and the refinery there will receive crude from offshore production platforms. Participants in the enterprise will include oil companies from India and the US, and the refinery is expected to come on stream in 2009. About half of its products will be for domestic use, the other half for export.

London Array
Name given to a proposed offshore wind farm to provide electricity for London and the southeast counties of England. The planned capacity is 1 GW and it is intended to site the wind farm in the Thames Estuary over a 245 km² area. One of the participating organisations is Shell.

London Underground, electricity consumption by
The electricity of the London Underground (LU) is estimated to be about 150 MW. The LU has gas turbine generators at Greenwich, south of the Thames, which use distillate fuel oil and natural gas to generate up to 105 MW. The deficit is bought from electricity supply companies.

Long Island Wind Park
A projected offshore wind park close to Long Island which will feature 40 turbines. It will be off the east side and therefore in the Atlantic Ocean, not in Long Island Sound.

Long string

The last string of casing in an oil or gas recovery well, just above or through the reservoir production zone. It is the longest casing in a well and has the smallest diameter.

Los Angeles Basin

Once the scene of major oil endeavours, where there was not only production but also downstream activity as early as 1910. It is estimated that to date there have been 8.5 billion **bbl** obtained from the oil fields of the Los Angeles Basin. At the present time, there are 43 active fields producing 87 000 bbl per day of oil. Some of the fields previously having been declared no longer economically workable have come back into service.

Louisiana Offshore Oil Port (LOOP)

Situated close to the Louisiana coast. A tanker (usually a **supertanker**), arriving at LOOP is moored to one of three buoys (single-point mooring bases) at the facility. Crude from the tanker is transferred to a subsea pipeline and from there to a marine terminal. From there the oil is transferred to shore and stored in a space capable of holding up to 48 million **bbl** of oil. This space comprises natural **salt domes**. From there the oil is piped to refineries both in Louisiana and in other states.

The facility was closed down for a number of days as a result of **Hurricane Katrina**, re-entering service in early September 2005.

Low-grading

The practice of extracting resources of the lowest quality first in order to improve the quality of subsequent stocks (see **high-grading**).

Lower 48 States landmass, proportion of occupied by coal

The most reliable estimate to date is that 13% of the landmass of the Lower 48 States of the US has coal underneath it.

Low NOx burners

Conventional pulverised coal **burners** have been developed over the years to give both rapid mixing of pulverised coal and air, and high flame temperatures with a view to achieving both high intensity of combustion and high combustion efficiency. Unfortunately, these conditions also favour **NOx** formation. Development of low NOx burners commenced in the early 1980s, with attempts to reduce both temperature and oxygen availability in the flame by various means. Staged supply of air with swirl to the burner was found to be an effective

method for NOx reduction and the first generation of low-NOx burners was developed on this principle.

The most recent low NOx burners (or ultra-low NOx burners) are
low-swirl burners being developed to emit less than 5 volumetric parts
per million NOx without any efficiency penalties. The burner achieves
ultra-low NOx emissions by merging technology advancements in
advanced lean premixed burners and fuel pre-treatment. The resulting system combines a low-swirl flame stabilisation method with internal flue gas recirculation.

LPG
See **Liquefied petroleum gas**

Lucapa-1
Newly developed oil well in Angolan waters, at a water depth of 1200
m and awaiting appraisal. Total have a 20% holding in the enterprise.
The well is in Block 14 of the grid system which applies in Angolan
offshore activity and represents the tenth major discovery in Block 14
over the last decade.

Lukasiewicz, Ignacy
Rival to **E L Drake** in the claim to have sunk the first oil well.
Lukasiewicz was a Polish apothecary whose initial contribution to oil
matters was distillation of crude oil having seeped out of the ground
naturally to make an illuminating oil; he registered a patent of this in
1853. Orders for such oil followed, and Lukasiewicz and two associ-
ates established wells up to 50 m in depth in the Gorlice region of
Poland. By the late 1850s other developers had participated and wells
as deep as 150 m were yielding crude oil.

Lukoil
Russian oil company, accounting for about 19% of the total Russian
oil production, having expanded and established a presence in other
countries including Egypt and Cyprus. Lukoil currently owns 28 retail gasoline outlets in Cyprus which had previously been owned by
either BP Cyprus or Exxon Mobil Cyprus. The company is also strongly
represented in some of the former eastern bloc countries including
Romania and Bulgaria.

Lurgi process
A means of **gasification** of coal having close similarity to **blue water**

gas manufacture, that is, reaction of a hot bed of coal with steam, but with the difference of a very high reacting pressure. An operating pressure of 20 bar is typical and this has the advantage of enhancing the yield of methane. A gas produced by the Lurgi process has up to 20% methane. Other flammable constituents are those of blue water gas: **hydrogen** and **carbon monoxide**. A gas produced by the Lurgi process has a calorific value of about 16 MJ m^{-3}, about halfway between that of blue water gas and that of **retort coal gas** or coke oven gas.

Although the Lurgi process is undoubtedly a total rather than a partial gasification method, with a high-volatile starting material such as **brown coal** there can be a small amount of by-product organic liquid called Lurgi tars.

Maari field

New Zealand offshore oil field, reserves 50 million **bbl**, where production is expected to begin in 2008 by means of a **floating production, storage and offloading**. The useful life of the field is estimated as only about ten years.

Maceral

Term first coined by Dr Marie Stopes (1880–1958), meaning the organic analogue of a mineral. The etymological basis is that to macerate something is to soften it by soaking it in water, which is what happens at the commencement of **coalification**. Unlike a mineral (e.g. silica, SiO_2) a maceral has no precise chemical composition. Classification of macerals is to some extent subjective, and is under continual international review, but there are three groups: **vitrinite**, **liptinite** and **inertinite**. Macerals within a particular coal sample can, to some extent, be separated by density methods.

Mackenzie Delta

This location in northern Canada contains some of the most concentrated reserves yet discovered of **natural gas hydrates**. There is current exploration there by the Japanese National Oil Company who are performing similar operations at the **Nankai trough**. Canada and north western USA are seen as being promising sites for eventual exploitation of natural gas hydrate reserves, there also being significant activity at **Northern Cascadia** and off the Oregon coast.

Mackenzie pipeline
Projected pipeline to carry natural gas from the Canadian Arctic to Alberta along a distance of about 800 miles, construction to take place over four winter seasons. A native tribal group, the Deh Cho First Nation, are currently opposing the building of the pipeline and have filed a suit in federal court to prevent the project from being started.

Madagascar, energy scene in
This African island country produces no oil or gas of its own. It utilises its quite abundant reserves of **hard coal** only at a level of 10^4 tonnes per year. It imports about 2.5 million **bbl** of crude oil per year which is processed in a local refinery. There is much use of wood waste and, in the sugar industry, of **bagasse**. There are seven hydroelectric installations, which between them provide two-thirds of the country's electricity. The total electricity generating capacity (hydro and thermal) is less than 1 MW. Domestic heating sometimes consists of wood waste or coal in a brazier.

Mae Moh
A power plant in northern Thailand producing > 2500 MW using locally won **lignite** as fuel. The lignite from the region not used in power generation is diverted to other industries including cement manufacture. Thailand imports large quantities of **hard coal**.

Maersk Riga
A **double-hulled oil tanker**, built in China with Monrovia, Liberia, as its home port. It makes about 30 voyages in a year, relocating about 6 million **bbl** of oil.

Magnesia
Mineral substance and the basis of a **refractory material** in **combustion** plant. Manufacture of magnesia bricks involves firing an initially semi-solid hydrated form of magnesia at about 1500°C. The bricks formed in this way have softening temperatures of about 2000°C, significantly higher than for most such materials based on fireclay.

Magnus field
The most northerly oilfield in UK waters, discovered in 1974 and operating ten years later; current production is 67 000 **bbl** per day. **En-**

hanced oil recovery is underway, oil so produced being taken to **Sullom Voe**. Following the enhanced oil recovery measures the field is expected to be productive until 2015.

Magnetohydrodynamic (MHD) generation of electricity
The generation of electrical power from ions in flames, enabling a combustion process to produce such power directly. Ions naturally present in flames include H_3O^+, HCO^+ and CH^+ and the ion concentration can be raised by use of an additive, possibly potassium or a halogen. A flame so enriched in ions is burnt between the poles of a magnet, resulting in an electrical potential.

There was much development work in the 60s. In April 1971, a 25 MW MHD generator using a natural gas flame came into operation in the then Soviet Union, and in the same month a smaller unit began operating in Germany. However, the expectations of the developers of the contribution of MHD to world power supply were unfulfilled.

Mahogany project
Name assigned to an investigative undertaking by Shell into *in situ* retorting and processing of Colorado Shale. The oil shale deposits of Colorado, Utah and Wyoming have more oil than all of the Middle East combined. Fuel from this resource would need to be retorted, hydrogenated and desulphurised to give an equivalent of conventional crude oil. The **EROEI** figure for the process will not be known before 2010. Even if it is greater than one, there might be difficulties in attracting investors as **Black Sunday** is still well within living memory.

Malawi, Electricity Supply Commission of
Unreliability of electrical power has in the past retarded progress in this southern African country, a situation which the statutory electricity commission has sought to improve. The country has a capacity of 224 MW, 95% hydroelectric. The remaining 5% is from **gensets**. There is a move to make a provision for the import of electricity from neighbouring countries such Mozambique.

Malaysia, export of fuel wood from
Quality fuel wood is an exportable commodity, there being about 40 Mt annually exported from Malaysia. The major markets include the UAE and Saudi Arabia, Hong Kong, Japan and Thailand. The part of Malaysia most productive of fuel wood is Sarawak. Sabah produces only a very limited amount specifically for the South Korean Market.

Malta, energy scene in

Malta, energy scene in

Malta imports oil from Libya. About 60% of it is used to make electricity, generation having been at 595 MW in 2005. There has been limited exploration for oil and gas in Maltese waters but with disappointing results. However, current high oil prices have renewed incentive for exploration and Malta is hoping to interest companies including Occidental in obtaining exploration licences.

Mangala field

Onshore oil field in India operated by Cairn Energy, UK, and expected to be producing 100 000 **bbl** per day by 2008. About 70% of this will be refined by the state owned Oil and Natural Gas Corporation (ONGC).

Mangrove

A tropical shrub used as a cellulosic fuel in some parts of the Far East.

Marathon Detroit refinery

This refinery in the so-called automobile capital of the world has recently been expanded from 74 000 to 100 000 **bbl** per day capacity. In addition to the refinery at **Garyville LA**, Marathon have refineries in Ohio, Kentucky, Illinois, Texas and Minnesota. Their total capacity approaches a million bbl per day, about 1% of the world's total refining capacity.

Marcasite

A form of iron sulphide FeS_2, therefore having the same chemical formula as **pyrite**, but a different crystalline form. It occurs as a mineral in some coals.

Marcinelle, Belgium

Scene of a **coal** mining accident in 1956, in which 262 lives were lost. Over 100 of the dead were migrant workers from Italy.

Marine current energy

Production of electricity from the kinetic energy of seawater which turns the blades of a submerged turbine. The operation is entirely analogous to that of the **wind turbine**. Development work is currently focused on island or coastal settlements lacking supply from the grid.

Marine incursion

Entry of the sea into a deposit undergoing **coalification**, later to withdraw. The site and extent of such incursions can be identified by a very significant difference in mineral content between two beds of coal adjacent to each other and similar in petrographic composition.

Marine wave energy

Production of electrical energy from sea waves. The principle differs from **hydroelectricity** generation and **marine current energy** in that in marine wave power generation it is air, not water, which turns the turbine blades. The air enters the turbine from a device called an **oscillating water column**. There has been a revival of interest in power from marine wave energy since the **Kyoto protocol** and at the present time there is significant addition to the national grid from power so generated at the island of **Islay**.

Announcements were made recently for the world's largest wave farm to be built off the coast of Orkney, Scotland in 2008. It will comprise four wave energy converters (known as Pelamis), each 520 feet long and generating 750 kW.

Mars platform

The largest of the Gulf coast offshore platforms to have been affected by **Hurricane Katrina.** Operated by Shell (71.5 %) and BP (28.5 %), Mars previously accounted for about 5 % of the total gas production from the Gulf. Repairs required following Katrina were extensive and involved a first for industry. Pipelines at depths of 2700 ft below the sea were repaired *in situ* by robots where, in shallower waters, divers would have been used.

Marsh gas

A gas composed principally of methane (CH_4) which is colourless, odourless and tasteless. Produced in marshlands by decaying organic matter.

Mauritius, energy scene in

This small country in the Indian Ocean has no known oil reserves and so imports crude oil. The country grows sugar and the **bagasse** residue finds fuel application although, unlike other sugar producing nations such as Brazil, Mauritius has not made **ethanol** available as a motor fuel. There is **hydroelectricity** in Mauritius although it is not sufficient to meet the country's entire power needs.

A new energy development in Mauritius is coal/bagasse co-firing to generate electricity at about 70 MW and also heat, in a **combined heat and power** cycle. In addition to the intrinsic advantages of co-firing, the installation also has flexibility to respond to seasonal variations in the availability of bagasse. At certain times of the year, coal will be the dominant fuel and heat from the cycle is returned to the sugar mills for use there. (Previously, such heat was obtained simply by direct combustion of the bagasse.) At other times, the primary fuel will be bagasse. Over the year, about a third of the power will be from bagasse combustion, reducing the coal import requirement very significantly.

Mawson Research Station Antarctica, energy supply at
Until recently, the Station was entirely dependent upon **diesel** brought from Australia, but now has two **wind turbines** with a third awaiting installation. The Station's power needs are about 600 kW, over two-thirds of which will be supplied as wind power when the three turbines are working.

MBDOE
See **Million barrels per day of oil equivalent**

McAfee cracking process
With the support of the Gulf Refining Company, Almer McDuffie McAfee (1886–1972) developed the petroleum industry's first commercially viable **catalytic cracking** process, a method that could double or even triple the gasoline produced from crude oil by standard distillation methods at that time. Based partly on an 1877 Friedel-Crafts patent, the McAfee cracking process required anhydrous aluminium chloride, a catalyst that was prohibitively expensive. In 1923, McAfee and Gulf developed a way to synthesise the catalytic reagent at low cost on an industrial scale.

Medium-density fibreboard (MDF), fuel use of
In County Antrim, Northern Ireland, a small company manufacturing this product has become self-sufficient in terms of electrical power by using the waste as a fuel to raise steam which undergoes a **Rankine cycle**. The waste comes largely from sanding the fibreboard, and power generation is up to 165 kW. The fluid at the conclusion of the cycle has sufficient **quality of heat** for hot water to be provided as a bonus, via a heat exchanger.

Membrane-electrode assembly
The combination of anode-membrane-cathode used in **polymer electrolyte membrane fuel cell**s. The first membrane-electrode assemblies

were constructed in the 1960s for the Gemini space program, and used 4 mg of platinum per cm^2 of membrane area. Current technology varies with manufacturer, but total platinum loading has decreased to about 0.5 mg cm^{-2}. Laboratory research now uses platinum loadings of 0.15 mg cm^{-2}. This corresponds to an improvement in fuel cell performance since the Gemini program, as measured by amperes of current produced, from about 0.5–15 A per milligram of platinum.

The polymer electrolyte membrane is a solid, organic polymer, usually poly (perfluorosulphonic) acid. A typical membrane material consists of three regions:

• the Teflon-like, fluorocarbon backbone, hundreds of repeating – CF_2–CF–CF_2– units in length

• the side chains, –O–CF_2–CF–O–CF_2–CF_2–, which connect the molecular backbone to the third region

• the ion clusters consisting of sulphonic acid ions, $SO_3^-H^+$.

The negative ions, SO_3^-, are permanently attached to the side chain and cannot move. However, when the membrane becomes hydrated by absorbing water, the hydrogen ions become mobile. Ion movement occurs by protons bonded to water molecules, hopping from one SO_3^- site to another SO_3^- site within the membrane. It is this mechanism that makes the solid hydrated electrolyte an excellent conductor of hydrogen ions.

The thickness of the membrane in a membrane-electrode assembly can vary with the type of membrane. The thickness of the catalyst layers depends upon how much platinum is used in each electrode. For catalyst layers containing about 0.15 mg of Pt cm^{-2}, the thickness of the catalyst layer is close to 10 mm, less than half the thickness of a sheet of paper. This membrane-electrode assembly, with a total thickness of about 200 μm (0.2 mm), can generate more than half an ampere of current for every cm^2 of membrane-electrode assembly at a voltage between the cathode and anode of 0.7 volts.

Membrane insulation, liquefied natural gas (LNG)-bearing vessels

Until recently, most **LNG**-bearing vessels had spherical tanks containing the LNG which were thermally insulated; balsa wood is a common choice of insulator for this application. This has the disadvantage that most of the interior space of the ship is unoccupied, the payload being mainly above the waterline. If instead the vessel has membrane insulation, the tank containing the LNG can be integral to the hull of the ship with an outer and an inner membrane. The space between the membranes is evacuated so that heat transfer to the LNG is prevented by the very high convective thermal resistance of the low-pressure contents of the space in the membrane. This is reinforced by some thermal insulation on the surface of the outer membrane. The advantage is that the ship's own inherent cargo space is being used to carry the payload, instead of spherical tanks which are an addition to the ship's basic structure.

Among the shipbuilders now manufacturing such vessels is the French ALSTOM Marine, which supplied its first LNG tanker in 1971—*Descartes*—which is still in service. The company, like its competitors, installed spherical tanks at that time but currently over 80% of the orders for LNG vessels are for the membrane type.

The membrane concept for LNG storage is also being used for vehicles e.g. cars, buses and trucks which run on LNG. It is also used in tankers for LNG transportation by road or rail.

Merchant incineration
Incineration of solid waste on a commercial basis, the charge being approximately according to weight. The incinerator operators accept different sorts of waste e.g. **municipal solid waste**, plastic waste, rubber waste and textile waste. The conditions of burning, such as the air supply rate, are adjusted according to the nature of the waste, so that the maximum extent of combustion is achieved. **Emission standards** of atmospheric pollutants apply.

Mercury analysis
Mercury is sometimes present in fuels such as natural gas. Although the concentrations are minute, the levels need to be monitored nevertheless due to their toxicity. Mercury levels in natural gas as low as 0.001 mg m^{-3} can be measured with current instruments which first collect the mercury on the internal surface of a gold coil.

Mercury density
Density of a coal as measured by displacement of mercury. Mercury has a surface tension which precludes its entry even into the very largest pores of the coal. The mercury density and the **helium density** together enable the porosity of a coal—the proportion of the apparent volume occupied by voids—to be calculated.

Messoyakha field
Initially a conventional gas field, operating as such from 1970–78. The field did not deplete as expected over this period however, and it is

now believed that it was being replenished by a layer of **natural gas hydrates** underneath it, of which the operators were unaware.

Metal hydride storage
The ability of metals to absorb large quantities of **hydrogen** at relatively low pressures makes them ideal candidates for hydrogen storage reservoirs. In the metal matrix, the hydrogen atom interacts with the metal atoms and the electrons.

$$M + n/2\ H_2 \leftrightarrow MH_n + Heat$$

As shown by the equation above, heat must be added to release the hydrogen from the metal hydride phase. It is therefore possible to bring the hydrogen atoms very close together, much more so than even in liquid hydrogen. However, the weight of the metal matrix is substantially greater than the stored hydrogen; just 1% of the total mass of the metal hydride is hydrogen. Nevertheless, the hydrogen density in metal hydrides is significantly greater than for gaseous or liquid hydrogen.

Metallurgical coal
A coking or pulverised coal used in metals and alloys manufacture.

Metallurgical coke
A low-sulphur coke used for iron ore smelting. Often referred to as 'met-coke'.

Methane Arctic
One of a number of **liquefied natural gas (LNG)** ocean tankers owned by British Gas. Another is the *Methane Polar*. Each has been in service for about 35 years and currently displays the flag of Singapore. In addition, the company charters a number of LNG vessels including the well-known *Methane Princess*. British Gas have also recently taken delivery of a new vessel, the *Methane Kari Elin*, built by Samsung. Its payload is > 60 000 tonne of LNG.

Methane Pioneer
This was the first vessel to convey **liquefied natural gas (LNG)** by sea. In 1959 it carried a cargo of LNG from Louisiana to Canvey Island (where there is still LNG storage), which is in the Thames estuary, London. The vessel was not purpose built—it had been a freighter in World War II. The *Methane Pioneer* voyage had demonstrated that LNG can be carried safely on the ocean. Significant trade of the substance between nations began soon afterwards, initially between Algeria and the UK.

Methanol

An organic compound with the formula CH_3OH. It is of course combustible, with a calorific value of 23 MJ kg^{-1}, about half that of gasoline. Its reactivity makes it suitable for use in spark-ignition engines, in pure form or blended with gasoline. It can be manufactured from coal via **synthesis gas** or by partial oxidation of natural gas.

Methyl tertiary-butyl ether (MTBE)

Used as an additive for gasoline. Chemical structure:

$$H_3C-O-\overset{\displaystyle CH_3}{\underset{\displaystyle CH_3}{|}}-CH_3$$

However, there is the difficulty that MTBE leaking and finding its way into the water supply will cause contamination. MTBE in water at a level of parts per billion can have such an effect, and in the US there has been lobbying by environmentalists to have MTBE replaced by ethanol as a gasoline additive. A severe case of MTBE contamination of water occurred in Santa Monica California about six years ago. MTBE had leaked from underground storage tanks into the water supply and several oil companies were ordered by the environmental protection agency to provide the residents of Santa Monica with an alternative supply of fresh water.

Mexico, geothermal power generation in

Currently Mexico generates about 1 GW from geothermal installations at locations including La Primavera and Los Azufres. There is also significant activity in Baja California where a new **geothermal electricity** plant recently began production at Las Tres Virgenes.

Micro-emulsions

Term applied to **biodiesel** fuels prepared by emulsifying **soybean** oil or palm with **ethanol**, with a higher alcohol as a stabiliser. Promising results have been obtained in compression-ignition engine trials with such fuels during tests conducted in Germany.

Microturbine

A turbine operating according to the **Brayton cycle**, therefore classifiable as a gas turbine, but having a sub-MW output. Microturbines are currently under development and are seen as having promise as mobile electrical power plant and very possibly as power plant for vehicles.

Middlings
A term used specifically for coal particles that contain both inorganic and organic material. The term is also used more generally to describe the intermediate grade or quality materials, for example of an ore.

Million barrels per day of oil equivalent (MBDOE)
A unit where energy from oil and other sources, including natural gas and coal, can be expressed in aggregate, for example in **world oil and gas consumption**. The unit is also widely used in estimates of future energy needs.

Millstone Power Station, Connecticut
The power station has two nuclear reactors. The first ran from 1970 to 1998 until decommissioning began, and the second from 1975 to the present time, ownership having changed in 2001.

Millstone Power Station has been under investigation by the US Nuclear Regulatory Commission after two **fuel rods** were unaccounted for in an inventory check in 2000.

Mina al-ahmadi
The location of a refinery, close to Kuwait City. It was built in 1949 with only local needs in mind, having at that time a capacity of 25 000 **bbl** per day. The refinery was significantly enlarged in the 80s and the original capacity increased by about a factor of 10–15. One of its important roles is the provision of fuel for oil tankers departing Kuwait, and it also supplies gasoline and kerosene for the Kuwait domestic market. Repairs to parts of the refinery had to be effected after damage sustained in the Gulf War.

Minemouth power plant
Electrical power plant built close to the pit-head of a coalmine to minimise transportation costs and guarantee security of supply. The downside is experienced when the mine closes. For example, the closure of the Selby coalfield in Yorkshire has meant that **Drax**, Eggborough and Ferrybridge power stations, all now inconveniently in-land, have to bring coal from the ports by train or road.

Minerals and inorganics, distinction between in coal

Some coal chemists make a distinction between minerals and inorganics on the following basis. Minerals are substances such as silica, clay and **pyrite** which are not combined chemically with the organic structure of the coal whereas inorganics are metal ions, chiefly calcium and sodium, bonded to sites such as carboxylate groups within the coal structure. Inorganics are removable by ion exchange. The distinction is relevant primarily to **brown coal**. For coals of higher rank, the term mineral matter suffices without knowledge of what proportion, if any, is bonded to sites in the coal. The minerals and inorganics concept has in recent years been extended to wood.

Minerva

(1) Coal field in Queensland, Australia, where production began as recently as May 2005. The coal is bituminous in rank and reserves are estimated as 26.8 Mt which, at the projected rate of production once operations are underway, will give the mine a life expectancy of 11 years. It is all destined for export to Asian markets.

(2) Gas field in Australian waters, a shallow field with drilling at only 60 m. It produces 4 million m³ per day with an accompanying 500 **bbl** of **condensate**.

Miscanthus

Type of grass and an **energy crop** currently being grown and utilised in places such as Scotland. It differs from the **short rotation coppice** in that a crop of *Miscanthus* is harvested in total annually, whereas only parts of a tree in a coppice are removed every 2–5 years and the tree itself remains, yielding fuel wood intermittently for up to 30 years. Of course, *Miscanthus* is only harvested once established on suitable land, when it yields an annual crop about 15 times.

In botanical terms, *Miscanthus* is a genus, there being several species. It originates from Japan and Korea but will grow in varied climates.

Mittelplate

The sole oil-producing field in the German sector of the North Sea. The Schwedeneck-See field, also in German waters, is no longer viable and decommissioning is taking place there. Most of Germany's gas and oil reserves are onshore.

Mixed oxide (nuclear) fuel (MOX)

When a nuclear reactor requires fuel assembly replacement it contains about 1% of plutonium, a transuranic element formed by neutron capture by U^{235} followed by β particle emission, having the potential to decay and release heat in so doing. The plutonium can be separated and, in the form of its oxide, blended with depleted uranium oxide from the isotopic enrichment to form MOX which is itself used as a nuclear fuel in electricity generation. Like uranium fuel itself it is pelletised and installed in **fuel rods**.

There is significant MOX production and usage in France and Belgium. There are proposals for its use in Japan, the spent waste from Japan being sent to Europe for conversion to MOX and then returned.

Mobil Delvac

Delvac = $\underline{d}$iesel $\underline{e}$ngine $\underline{l}$ubricants $\underline{vac}$uum. The term vacuum derives from the vacuum distillation of petroleum material required for lubricant manufacture. Formed in 1925, Mobil Delvac has engaged in R&D continually. A 'plus' for Mobil Delvac was the appearance at a recent truck show in Louisville KY of a diesel powered Volvo truck which had over a seven-year period covered a million miles. Experts commented on the remarkably good condition of the engine, which over the million miles covered had used no lubricants other than those manufactured by Mobil Delvac.

Molecular seal

Device which restricts air ingress at a **burner** in the event of flame extinction, reducing the risk of explosion within the burner pipe work.

Molten carbonate fuel cell

Molten carbonate fuel cells use an electrolyte that conducts carbonate (CO_3^{2-}) ions from the cathode to the anode. The electrolyte is composed of a molten mixture of lithium and potassium carbonates, retained by capillary forces within a ceramic support matrix of lithium aluminate. At the fuel cell operating temperature, the electrolyte structure is a thick paste, and the paste provides gas seals at the cell edges. Molten carbonate fuel cells operate at about 650°C and a pressure of 15–150 psig (1–10 barg). Each cell can produce up to between 0.7–1.0 V DC.

The advantages of molten carbonate fuel cells are that they:

• support spontaneous internal reforming of light hydro-carbon fuels
• generate high-grade waste heat
• have fast reaction kinetics (react quickly)
• have high efficiency, and

- do not need noble metal catalysts.

The disadvantages are that they:

- require the development of suitable materials that are resistant to corrosion, are dimensionally stable, have high endurance and lend themselves to fabrication
- have a high intolerance to sulphur. The anode in particular cannot tolerate more than 1–5 ppm of sulphur compounds (primarily H_2S and COS) in the fuel gas without suffering a significant performance loss
- have a liquid electrolyte, which introduces liquid handling problems, and
- require a considerable warm-up period.

Molten carbonate fuel cells can operate using pure hydrogen or light hydrocarbon fuels. When a hydrocarbon such as methane is introduced to the anode in the presence of water, it absorbs heat and undergoes a steam reforming reaction:

$$CH_4 + H_2O \rightarrow 3H_2 + CO$$

When using other light hydrocarbon fuels, the number of **hydrogen** and **carbon monoxide** molecules may change but in principle the same products result.

The reactions at the anode are:

$$3H_2 + 3CO_3^{2-} \rightarrow 3H_2O + 3CO_2 + 6e^-$$
$$CO + CO_3^{2-} \rightarrow 2CO_2 + 2e^-$$

The reaction at the cathode is:

$$2O_2 + 4CO_2 + 8e^- \rightarrow 4CO_3^{2-}$$

The CO_3^{2-} ion is drawn through the electrolyte from the cathode to the anode by the reactive attraction of hydrogen and carbon monoxide to oxygen, while electrons are forced through an external circuit from the anode to the cathode. Combining the anode and cathode reactions, the overall cell reactions are:

$$2H_2 + O_2 \rightarrow 2H_2O$$
$$CO + \tfrac{1}{2}O_2 \rightarrow CO_2$$

The fuel cell therefore produces water regardless of fuel, and carbon dioxide if using a **hydrocarbon** fuel. Both products water and carbon dioxide must be continually removed from the cathode to facilitate further reaction.

Monoethanolamine

Organic compound used for the removal of carbon dioxide from hydrocarbon gas. Carbon dioxide so removed can be used in **enhanced oil recovery**. The structural formula of monoethanolamine is:

$$
\begin{array}{c}
H \\
\backslash \\
N - CH_2 \\
/ \qquad \backslash \\
H \qquad OH
\end{array}
$$

Its role in carbon dioxide removal is that of a solvent. However, monoethanolamine is used as a reagent in the removal of sulphur from liquid hydrocarbons, the organic sulphur in the fuel having first been converted to hydrogen sulphide which then reacts with the monoethanolamine.

Monopropellant

A fuel capable of heat-release without external oxygen supply. An application is to **rocket** propulsion; hydrogen peroxide is a common example.

Morecambe Bay gas field

A gas field in the Irish Sea, off the west coast of Britain. The reserve was discovered in 1974 and production began a few years later. There was a fatal accident involving a helicopter close to the field in late 2006.

Mossel Bay, Cape Province, South Africa

The location of four 150 MW gas turbines for power generation, a response primarily to rising domestic rather than industrial demand. The use of gas turbines instead of steam turbines in long-term supply to residential users is fairly unusual.

Motiva Enterprises

A joint enterprise of Shell and Saudi Refining Inc. and operator of three refineries in the Gulf States including one at **Port Arthur, TX**. Expansion of these is planned with a view to adding between 100 000–300 000 **bbl** to the daily quantity of crude processed.

Moura

Locality in central Queensland, Australia, the scene of a newly discovered mine capable of producing 12 Mt yr^{-1} of coking coal over a

40-year period. Production is expected to begin in 2011. One major beneficiary is likely to be the Australian steel industry. There are already established coalfields in central Queensland, including that at Blair Athol.

Mpumalanga Province
Part of the eastern Transvaal and the location of **bituminous coal** production until 1993. Some of the coal produced there was used locally while some was taken 560 km along a dedicated railway track to **Richards Bay** for export. This was additional to the coal from Kwazulu-Natal, which is exported via **Richards Bay**.

MSW
See **Municipal solid waste**

MTBE
See **Methyl tertiary-butyl ether**

Mud
Colloquial term for **drilling fluid**.

Mud ring
If a particularly porous part of the geological formation is encountered during rotary drilling, there will be loss of the aqueous component of the **drilling fluid** and consequent concentration and deposition of the remaining material to form what is known as a mud ring.

Mumbai, liquefied natural gas reception at
There is a **liquefied natural gas** terminal at Mumbai (formerly Bombay), India, currently capable of receiving 5 Mt yr^{-1} of the substance. Expansion is expected to enable the facility to receive 10 Mt yr^{-1}. The terminal is owned and operated by **Petronet**.

Municipal solid waste (MSW)
Waste from households having some potential as a fuel if processed by drying, removal of non-combustible constituents and (possibly) pelletising. Once described by the fuel technologist N Y Kirov as an 'energy ore', it has in its raw state a calorific value of 7–8 MJ kg^{-1} which (as Kirov points out) is equivalent to that of many low-rank coals when bed moist. MSW has limited usage and its applications include steam raising. Difficulties in its use include its heterogeneous composition and that manual handling of it is unhygienic.

Interestingly, opinions differ as to whether MSW is a renewable source of energy. There is the view that much of the combustible component is plastic, derived from oil and therefore not renewable. In the US, the Federal Government and some state governments classify MSW as a renewable fuel on the basis that much of it is biomass, which is of course renewable. This is not of course simply a fine point of semantics: allocation of emission allowances, for example, will depend on whether the fuel is in the renewable or non-renewable category.

As an example of the fuel potential of MSW, let it be noted that the Netherlands produce 325 MW of power for the grid with MSW as fuel.

Murdoch, W
The developer, in the late 18th Century, of the process for preparing an illuminating gas from coal. He was employed by Boulton and Watt, Birmingham, England, at the time.

Murmansk Shipping Company
Currently the world's largest operator of nuclear powered vessels, although the entire fleet of eight such vessels is in fact state owned. Most of the vessels are ice-breakers for use in the Arctic region.

Muskeg River mine
Tar sand mine in Alberta having come into production only in the last few years. It has a 30-year life expectancy and it is predicted that it will yield 1.65 billion **bbl** of **bitumen** over this period.

Myitnge River, Myanmar
The location of a proposed **hydroelectric** generator, about 50 km from the major city of Mandalay. Power will be generated at 790 MW level for transmission to the whole of Myanmar.

N

Nakhodka
(1) Port town on the east coast of the former Soviet Union. A pipeline to convey crude oil from Russian fields to Nakhodka is planned, the oil being transported from Nakhodka to Japan or to the western USA.

(2) A Russian tanker which sank in Japanese waters in 1997 releasing 36 000 **bbl** of crude oil. Beaches and shellfish beds were threatened.

Namibia, prospects of oil off the coast of
Bordered by Angola, Namibia is seen as a possible future second to Angola in oil production in southern Africa. Exploration in Namibian waters is taking place and the Houston based company EnerGulf are currently at work in Namibian waters, using proprietary information supplied by oil companies who had carried out partial exploration of the area previously. It is believed that 500 million **bbl** of crude oil (a **giant field**) await discovery.

Nanjing
The site of a new **ethylene** manufacturing plant operated by **Sinopec**. It will produce 900 000 tonnes of ethylene per annum by **cracking**. The German multinational BASF have a major interest in the plant.

Nankai trough
Promontory in southwest Japan where there is exploratory drilling for

natural gas hydrates. Ever responsive to new sources of energy, Japan also has involvement in such exploration in the **Mackenzie Delta**.

Nanticoke Power Plant

Thermal power generation plant at the Canadian side of Lake Erie, operated by Ontario Power Generation and one of the largest such facilities in the world. It has eight 500 MW steam turbines giving a combined capacity of 4000 MW. Coal for the plant is received from diverse sources and blended.

Naphtha

Product of crude oil refining in the boiling range between gasoline and kerosene. Sometimes it is mildly chemically treated and blended with straight-run gasoline. A great deal of naphtha is cracked to produce alkenes for the petrochemical industry. It has also been used to make **town gas** by **steam reforming**.

Naphtha is also an obsolete term for coal tar.

Naphtha, use of in power generation

The **Dabhol project** will use naphtha initially until its facilities to receive **liquefied natural gas** are in place and there is use of naphtha as a fuel for power generation in other countries including Libya. However, such use is not widespread as naphtha attracts a high price as a source of organics for chemical processing, so burning it to generate electricity is under most circumstances wasteful: natural gas is much more economic.

Naphthalene

A white solid with a strong coal-tar odour, used as a starting material for many organic compounds including dyes, insecticides, plastics, solvents and mothballs.

naphthalene ($C_{10}H_8$)

Nariin Sukhait, southern Mongolia

Location of a newly-discovered deposit of **bituminous coal** known to contain at least 100 Mt. As the deposit is only 40 miles from the Mongolia-China border, some Nariin Sukhait coal will probably be exported to China for wider distribution by rail. This is timely in view of the recent closure of many Chinese mines on grounds of safety.

Natuna Sea

Indonesian waters northeast of Sumatra where there is significant **non-associated gas** production. It includes the Macan field, where there has been a recent discovery of gas and where appraisal is underway, and the Anoa field which is already producing gas. Existing infrastructure will enhance the viability of bringing the Macan field into production.

Natural bitumen

A bitumen or high-viscosity petroleum found in **bituminous**, oil or tar sands. Not recoverable by conventional methods.

Natural gas, carbon dioxide in

Not only does the presence of natural gas reduce the **calorific value**, it also adds to the greenhouse emissions when the gas is used. Removal is therefore desirable in the case of those reserves of natural gas with appreciable CO_2. Separation by condensation at high pressure followed by re-injection into the well is an approach often taken. In Japan, there is investigative work into using solvents for such separation. Gas from the **Rhum field** is fairly high in **carbon dioxide**.

Natural gas, conversion to hydrogen for power generation

Natural gas is of course a CO_2 producer. There is an exploratory scheme involving BP and Shell among other organisations whereby natural gas is converted by **steam reforming** to hydrogen which in turn is used to generate steam for power generation. The chemistry is simple and the technology classical:

$$CH_4 + H_2O \rightarrow CO + 3H_2$$
$$\frac{CO + H_2O \rightarrow H_2 + CO_2}{CH_4 + 2H_2O \rightarrow 4H_2 + CO_2}$$

The calculation below relates to the power station where trials are proposed. It will require 70 million ft^3 per day of methane to sustain the hydrogen supply which the power generation will require. After carbon dioxide removal the hydrogen will be liquefied and stored as a cryogen before 'regasification' and use in steam raising. The calculation (the non-stoichiometric equation represents yields) addresses the power produced by this amount of natural gas so used.

$$CH_4 + 2H_2O \rightarrow 4H_2$$

Now 70 million ft^3 $\equiv$ 1.9 million m^3 containing:

$$1.9 \times 10^6 \times 40 \text{ mol} = 7.6 \times 10^7 \text{ mol} \Rightarrow 3 \times 10^8 \text{ mol H}_2$$

Heat released by this amount of hydrogen

$$= 3 \times 10^8 \times 285 \times 10^3 \text{ J} = 8.6 \times 10^{13} \text{ J}$$

If the steam turbine operates at say 33% efficiency, electrical energy produced per day = 2.8×10^{13} J

Power = $\{2.8 \times 10^{13}/(24 \times 3600)\}$ W = 325 MW

The quantity of natural gas used in the above calculation is for generation at 325 MW, a fairly typical performance for a single steam turbine in power generation. The hydrogen will of course burn to water vapour only. The advantage of the process is that the carbon dioxide, whether removed by a membrane or by a chemical reagent, will have been captured before combustion.

Natural gas, decarbonisation of

Decarbonisation of natural gas with water yields hydrogen with carbon dioxide as a side product. This can be removed and stored underground leaving hydrogen for use as a fuel, therefore decarbonisation is equivalent to **carbon sequestration** in advance of combustion. The thermochemistry of the process is outlined in the box below.

Direct methane combustion:

$$CH_4 + 2O_2 \rightarrow CO_2 + 2H_2O(g)$$

803 kJ of heat released per mol CH_4 reacted

Conversion of methane to hydrogen:

$$CH_4 + 2H_2O(g) \rightarrow CO_2 + 4H_2$$

164 kJ of heat taken from the surroundings per mol CH_4 reacted

Combustion of the hydrogen:

$$4H_2 + 2O_2 \rightarrow 4H_2O$$

968 kJ of heat released for the reaction as written

Combining the latter two reactions:

$$CH_4 + 2O_2 \rightarrow CO_2 + 2H_2O(g)$$

$$(968 - 164) \text{ kJ released} = 804 \text{ kJ}$$

It is advantageous in energy terms to convert the methane to hydrogen in what is simply classical **synthesis gas** technology. Work of this nature is taking place in Scotland where the carbon dioxide is taken to sub-sea storage sites.

Natural gas, seasonal production of

The motivation for **LNG** was that it enables gas produced during the warmer months, when demand is lower, to be stored for the colder months with all the pricing benefits accruing from such an arrangement. An alternative is for some of the gas produced during the warmer months to be stored under the sea. This is a reality at the Kinsale field off the Irish Coast, where the US company Marathon store not only gas produced at their own platforms but also gas produced at platforms owned by 'third party' companies in this way. This enables gas to be made available, for example for electricity generation, as needed without intermediate liquefaction.

Natural gas and crude oil, correlation of prices of

Natural gas sometimes occurs with crude oil as **associated gas** and the major oil producing regions of the world including the North Sea also have non-associated gas fields. A combustion plant sometimes has **burners** which can use natural gas, fuel oil or both. One would therefore expect there to be a correlation between the prices of the different types of gas.

Correlations applied in the US use the West Texas Intermediate (WTI) as a **benchmark crude** and the **Henry Hub** spot price for natural gas. A barrel of crude oil is capable of releasing 5.8–5.9 million **British Thermal Units (BTU)** on burning, and rounding this up to 6 gives the correlation:

Oil price (WTI) = 6 × Natural gas price (Henry Hub)

known as the one-to-six rule. However, if the prices of oil and gas over the years are examined, it is noticed that the prices fit the following correlation more closely:

Oil price (WTI) = 10 × Natural gas price (Henry Hub)

known as the one-to-ten rule. In recent months, when oil prices have been rising very sharply, there has been a transition from the one-to-ten rule towards the one-to-six rule. In the last quarter of 2004 the WTI price for oil was about $US 50 per **bbl** and the Henry Hub spot price for natural gas about $US 6 per million BTU, giving a factor of 8.3, just over halfway between the one-to-six and one-to-ten rule. Since then, oil prices have approached $US 70 per bbl.

At the time of writing, the WTI crude oil price is ≈ $US 60 per bbl and the Henry Hub spot price for natural gas ≈ $US 13. This gives a value for the ratio of just under five, making natural gas more expensive than would be expected even on the one-to-six rule. This must be seen as an aberration due to the instability of oil prices.

Natural gas hydrates

In the terminology of chemistry these are 'clathrates': substances in which molecules of one compound have become enclosed in the lattice structure of another. In natural gas hydrates, methane molecules have become enclosed in ice. Some experts use the term 'non-stoichiometric compound' instead of clathrate; both are correct.

Natural gas hydrates occur in the Arctic region, on the sea floor and on all continental shelves at depths in excess of 600 m. Such a depth corresponds to a pressure of:

$$600 \text{ m} \times 9.81 \text{ m s}^{-2} \times 1000 \text{ kg m}^{-3} = 59 \text{ bar}$$

and pressures of this order are required for formation of the methane hydrates. Methane is highly concentrated in the hydrate. A unit volume of the hydrate on dissociation releases about 160 units of volume of the gas at 1 bar and 300 K. There is disagreement about the total quantity of natural gas hydrates in the world. Estimates of methane extractable from hydrates in the range 10^{15}–10^{19} m^3 have previously been quoted, although the most recent estimates are in the 10^{14}–10^{15} m^3 range. The 2000 estimate for world reserves of natural gas was 1.5×10^{11} m^3 so even the lowest estimate for the gas potentially obtainable from hydrates exceeds this by about three orders of magnitude.

Natural gas hydrates feature strongly in discussions of post-oil energy generation. The only proven way to extract gas from the hydrates at this time is to reduce the pressure of any methane gas with which the hydrates are in phase equilibrium, known as the depressurisation method. This readily brings about release of the methane from the ice structure in which it is enclosed. Evidently this approach can only be taken at sources where the hydrate and gaseous methane are in contact, which requires encapsulation of free gas in the hydrate deposit. Such depressurisation occurred naturally when, unbeknown to the operators, a conventional gas field was on top of a hydrate reserve at the **Messoyakha field** in Russia.

Natural gas liquids (NGL)

NGL is composed of a mixture of ethane, propane, **butane** and possibly higher **hydrocarbons**, including pentane originating from natural gas. Although methane, the principal constituent of natural gas, cannot be made into a liquid by applying pressure at ordinary temperatures because of its low critical temperature, the minor constituents identified above can and are therefore separable from the gas in this way. Some reserves of natural gas are higher than others in these constituents; examples include the natural gas obtained offshore close to northwest Australia and the field at **Kapuni** in New Zealand. NGL will normally be stripped off and enter the liquid production stream, which is why some compilations of liquid fuel reserves for particular countries lump crude oil and natural gas liquids together. There are technologies for removing heavy components of natural gas by use of a membrane instead of by applying pressure.

The terms NGL and **condensate** often appear to be used interchangeably in the industry or are sometimes combined as NGL/condensate. The Norwegian sector of the North Sea is one of the world's largest producers of NGL and this terminology is used there. More conventionally, however, condensate means hydrocarbons of $\approx C_5$ which will condense naturally at room temperature and pressure. A reader should be aware that in the industry there is some variation in the use of these terms, and that some definitions of NGL encompass all of the hydrocarbon content other than methane, thereby eliminating the need for a separate word for the heavier components. A condensate field is one of **non-associated gas** with a high proportion of NGL/condensate, the **Britannia field** in the North Sea being a good example. By contrast, the **Morecambe Bay gas field** contains **dry gas**.

Natural gas pipeline, manufacture of

The usual material is mild steel and some such pipelines are seamless, but the majority are welded. In the seamless type a cylindrical block of steel is heated and a hole bored through it. In the welded type, a sheet of metal is curved to make a hollow cylinder which is welded. The weld is radiographed to ensure that it has the same design stress as the steel in the absence of a weld. (Such inspection of welds is standard practice. If the weld is not fully radiographed, a correction is ap-

plied.) If a pipe is to conform fully to an American Petroleum Institute (API) standard specifications apply not only to the pipe but also
to the steel from which it is made. Natural gas producers usually obtain pipe via a stockist rather than directly from the manufacturer.
Such stockists will buy surplus unused pipe from producers of natural
gas, even if it was not made by one of their usual suppliers, provided
that its conformity to a suitable standard is documented.

Natural pressure
The energy inherent in an oil or gas reservoir which causes the oil or
gas to rise to the surface unaided when a well is opened.

Negative financial value, fuels
Substances used as fuels that otherwise would have to be professionally
disposed of, incurring charges. Their purchase value is in that sense 'negative'. An obvious example is **municipal solid waste**. Another is spent cooking oil, which, after filtration, can be used as **straight vegetable oil**. Sometimes biomass fuels are so abundant as to have a negative value; this is
true of **rice husks** in at least one of the Central American countries.

Negawatt
A hypothetical tradeable unit of energy saved, introduced by Amory
Lovins in a 1989 speech. The negawatt is a method of increasing available electrical energy supply to consumers by conserving energy rather
than by increasing generation capacity. Negawatts can be 'generated',
for example, by temporarily shutting down air conditioning or by using only a subset of available lighting.

Nepal, deforestation threat to bird life
Nepal is a sanctuary for many species of bird; ornithologists have identified a total number of 831. It is therefore unfortunate that over 80%
of the Nepalese population use wood fuel, much of which has been
gathered without any regard to the condition of the forests or the
survival of the bird species which inhabit them. Of the 831 species,
133 are considered endangered, 72 of these cases to a critical degree.

Nerefco
See **Netherlands Refining Company**

Netherlands, biomass co-firing with coal in
This country aims at 12% replacement of coal by **biomass** in thermal

power generation. However, land is expensive in Holland and energy crop cultivation uneconomic. The biomass, which cannot be obtained in sufficient quantity from waste wood sources, might therefore need to be imported.

Netherlands Refining Company (Nerefco)
Nerefco operates a large refinery near Rotterdam in which Chevron have a 31% holding. It receives crude oil from 57 countries, some of it carried in **supertankers**. It also has a direct pipeline connection to one of the production platforms in the North Sea. Its refining capacity is 400 000 **bbl** per day. It supplies jet fuel to Schipol Airport in Amsterdam.

Neuquen
Province of Argentina and the location of major activity in **hydroelectricity**, current output being 1000 MW. Neuquen is also the site of despatch of natural gas from Argentina to Chile along the Gas Andes Pipeline, 288 miles in length.

Neutral zones
Also known as divided zones, areas where the border between two hydrocarbon-rich countries has not been agreed upon and a sharing arrangement of the resources within it is in operation. The best-known example is the Saudi-Kuwaiti Neutral Zone, which contains 5 billion **bbl** of crude oil to which Saudi Arabia and Kuwait, by an agreement formalised in 1971, currently have access on a 50–50 basis. Each country includes its half-share in the calculation of its own total oil reserves. For an interim period ending in 1983 there was such a zone between Saudi Arabia and Iraq. At the present time Saudi Arabia's border with the United Arab Emirates and that with Oman are still not settled. With offshore hydrocarbons there can also be disputes over maritime borders as is currently happening in the East China Sea.

New Brighton, PA
The location in October 2006 of a train derailment involving tank cars of **ethanol**, one of which broke open with a resulting fire and explosion. It is believed that this was the first time that ethanol had displayed **boiling liquid expanding vapour explosion (BLEVE)** behaviour. Transportation safety of ethanol in the US is the subject of considerable debate at the present time. Five billion US gallons of ethanol are transported by rail, road or (a small proportion) barge annually in the US, and there are calls for some of it to be conveyed by pipeline.

New England, wind power in
At present there is a wind farm in mountainous terrain in western Maine and proposals for another, comprising 30 more turbines, in the north of the State. Four German built **wind turbines** each with 1.5 MW capacity will be installed and commissioned in New Hampshire in 2007. Vermont has had wind power since 1997 at the Searsburg site, currently with 11 turbines which is expected to increase to 20 or more.

New York City incinerator
The site of the first recovery of heat from the burning of **municipal solid waste (MSW)** in 1898, at East 17th Street. There would have been a team of about thirty men and boys with the unappealing job of manually separating the combustible part of the MSW, chiefly paper, and putting it into sacks.

New Zealand, hydrocarbon reserves in
This country, which for very many years derived its prosperity from primary produce, is not well endowed with hydrocarbons. There is much natural gas at the offshore Maui field which contributes in two ways to the nation's transport fuel needs. First, natural gas can be steam reformed to produce **methanol** which, in a catalytic process, is made into petrol extender. Secondly, compressed natural gas (CNG) is itself being used directly as a transport fuel. There is onshore gas at **Kapuni**. In the same waters as the Maui field, oil was discovered in 2003 and again in 2004, to the extent of 27 million **bbl**. In what has become known as the Tui Area oil field development, it is planned that production will commence at 50 000 bbl per day in the second quarter of 2007. This will be New Zealand's first offshore oil installation. An offshore oil field off the east side of the North Island, the **Maari field**, is currently being developed and production there is expected to begin in 2008.

Newhall Refinery
Situated in the Santa Clarita Valley in California, this refinery was opened in 1884 to process oil from **Pico Canyon**. Although replaced by a new refinery in 1930, it was not demolished and is now a tourist attraction.

Newman Spinney, Scotland
The site of **underground gasification of coal** in the 1950s by the then National Coal Board. Although a technical success, it did not supply gas commercially.

Nicaragua, energy scene in

This country, the largest in Central America, consumes about a million tonnes of petroleum products, which have been refined in Nicaragua from imported crude, per year. Some export of refined material to other Central American countries takes place. Electricity generation is at ≈ 250 MW, of which one-quarter is hydroelectric and the remaining thermal using imported fuels.

Nickel-hydrogen battery

A hybrid combination of a **battery** with a **fuel cell**, it has a nickel oxide positive electrode similar to the nickel-cadmium cell, and is like the hydrogen-oxygen fuel cell since it has a **hydrogen** negative electrode. This hybrid battery has a long cycle life, high specific energy, high power density, and also exhibits tolerance for overcharge. It is therefore the battery of choice in many aerospace applications, especially geo-synchronous and low Earth-orbit satellites. In addition, the battery's hydrogen pressure is a good indicator of the charge state of the battery. Recently, nickel-hydrogen batteries have also been used in terrestrial applications. Its disadvantages include an expensive initial cost, as well as low volumetric energy density. Nickel-hydrogen batteries are bulky, require high-pressure steel canisters and cost thousands of pounds per cell.

Nigeria, effect on oil production of social instabilities in

In 2006 oil production in Nigeria dropped by 20% from the previous year causing financial losses of over $US4 billion, and this is attributed to the social milieu. A prevalent difficulty has been the seizing of foreign nationals at the oil fields as hostages. In January-February 2007 nine oil workers from China were kidnapped and held for 11 days.

Nigeria, oil production

There has been commercially significant production of crude oil in Nigeria for almost 50 years. Nigeria is currently the tenth largest oil producer in the world and the seventh largest exporter. There are 35 billion **bbl** of crude oil with promise of discovery of at least another 5 billion bbl over the next five years, plus 5×10^{12} m^3 of natural gas. Several foreign operators are active, the major company being Shell. Most of the crude oil is in the Niger River Delta and at present 75% of the associated gas is flared. This wasteful practice is being addressed by the installation of pipelines to take the gas to users, and the year 2010 has been set as the deadline 'zero flare'.

There are four oil refineries in Nigeria with a combined nominal capacity of just under 450 000 bbl per day. These have operated inefficiently due to factors including sub-standard management and corruption, therefore Nigeria has to import refined petroleum material. Some of the refined material produced in Nigeria is diverted to petrochemical manufacture.

Social and political instabilities have had a serious effect on Nigeria's oil industry. Environmental responsibility has not always been evident and damage from oil spills, excessive flaring of gas and the sabotaging of oil pipelines has been extensive. Legitimate complaints to the government have at best been ignored: at worst they have led to reprisals in the form of a death sentence.

Nigeria has been a member of **OPEC** since 1971.

Nitrated biodiesel
Spent vegetable based cooking oil nitrated with nitric acid has been shown, in exploratory work in Spain and in Greece, to be a good **cetane enhancer** for conventional (mineral) diesel. It has the advantage of being made from cheap starting materials (spent cooking oil is likely to have an EROEI of less than unity) and the possible advantage – further development work is needed – of being more chemically stable in storage than conventional cetane enhancers.

Nitrogen, use of in enhanced oil recovery
Nitrogen, if available, is as suitable as carbon dioxide for use in **enhanced oil recovery**. It is intended over the next decade to use nitrogen for this purpose at certain Mexican oil fields at an eventual rate of 2.5×10^6 m^3 of nitrogen per day. The nitrogen will be obtained by separation from air.

Nitrous oxide N$_2$O
Oxide of nitrogen, sometimes produced in combustion processes in particular in the combustion of wood. As a greenhouse gas it is 300 times as powerful as carbon dioxide. It has an atmospheric chemistry cycle independent of **NO$_x$**.

Non-asbestos materials
Also known as asbestos substitutes, these materials have replaced asbestos, which is widely known to be a threat to health. There are a wide variety of non-asbestos materials and a suitable one for a particular application can usually be found. Ingredients include carbon,

glass and a binder. The materials find extensive application in steam plant.

Non-associated gas
Natural gas that is not in contact with crude oil in a reservoir.

Non-caking coal
A coal that burns to leave an **ash** residue without fusing together or solidifying to form a **coke**.

Non-fossil fuel obligation
Scheme whereby a specified proportion of power supplied by each of the regional electricity companies in the UK must originate from sources other than the burning of fossil fuels. Such sources include **landfill gas**, marine waves, **photovoltaic cells** and wind farms. This requirement can be met in one of three ways: by ownership of non-fossil fuel generating plant by the electricity company itself; by purchase of non-fossil fuel power by a regional electricity company having no such plant itself from one which does; by purchase of non-fossil fuel power from an enterprise which produces only such electricity having no conventional generating plant of its own.

Electricity sold from a non-fossil fuel source to a regional electricity company attracts a premium price. The regional electricity company is reimbursed for the difference between that and the market price from revenue raised by the Fossil Fuels Levy which is of course paid by the electrical consumers themselves.

Unlike power stations using fossil fuels, a generating facility such as a wind farm or an assembly of photovoltaic cells can be dramatically impacted by the weather. For this reason a non-fossil fuel generator has to be able to demonstrate, before its bid with a regional electricity supplier can be considered, that the supply required of it would be well below its capacity under ideal conditions (that is, well below its **nameplate capacity**). Some simple calculations illustrating this follow.

> Typically, a wind farm will need to be of a capacity that it can supply at the agreed level to a regional electricity company if, because of unfavourable circumstances, its performance is down to ≈43% of that capacity. This means that for every MW which the wind farm has undertaken to supply, it must have a capacity of

$(1/0.43)$ MW = 2.3 MW

For photovoltaic cells, capability to deliver even if performance is down to $\approx 17\%$ of nameplate capacity must be demonstrated. Therefore, for every MW it is required to deliver, it must have a capacity of:

$(1/0.17)$ MW = 5.9 MW

Over-capitalisation in plant is therefore inevitable for the non-fossil fuel enterprise which accordingly has a statutory entitlement to the top price for any power it sells to a regional electricity company.

Non-recoverable
The portion of an oil or gas field that cannot economically be recovered.

Non-renewable energy
A source of energy, such as a **fossil fuel**, that has taken a longer period of time to form than the rate in which it is used and hence will become depleted at current consumption rates.

North America, electricity production in
In the US, electricity production was 435 GW in the year 2000. In Canada, there was an average load of 67 GW in the same year. A *per capita* consumption comparison is not possible from these figures because there are imports and exports between the two countries. For example, power generated at the **Nanticoke Power Plant** on Lake Erie goes partly to the Canadian Province of Ontario and partly to the US State of Ohio. There is of course distribution between different parts of the US. For example, in 2000 New York City produced at 8 GW and consumed at 10.3 GW so the deficit had to be brought from elsewhere.

Illustrative of the regional variations within the US is the fact that the State of Montana produces electricity at just under 3 GW. California produces at 32 GW. At present, 56% of the total US production of electricity is from coal. A little more than 0.5% is **geothermal electricity** and this sector is expanding.

North Carolina, possible offshore drilling
As yet, there has been no drilling off the coast of this State, which has a border to the south with South Carolina and to the north with Virginia.

It is however possible that by the 2020s there will be oil production in waters off North Carolina. Surveys about 40 miles out to sea have taken place and Chevron have estimated a 7% chance of finding oil and a 2% chance of finding it in commercially significant quantities. It is also believed that there are fields of non-associated natural gas.

North Dakota, lignite deposits in

Site of one of the world's most utilised **lignite** deposits. The deposit extends in the northern direction over the Canadian border and in the western direction over the state border with Montana. ND lignite currently sells in raw form for ≈$25 per ton, about two-thirds of the current **CAPP** price for a hard coal.

North Korea, nuclear activity in

There is always the difficulty that a country in possession of nuclear materials needs to satisfy the world that such materials are being used for civilian purposes, which usually means electricity generation, and not in the fabrication of nuclear weapons. In spite of agreement by North Korea to the denuclearisation of the country in talks held with other countries including China, the US and Japan, North Korea is asserting its rights to have a nuclear programme for power generation and there is in fact a uranium enrichment facility in North Korea. An inauspicious event was the expulsion by North Korea of UN weapons inspectors. However, recent talks led to hopes of an agreement whereby North Korea would receive energy aid in return for shutting down its nuclear programme, hopes which may yet be dashed.

North Korea, oil reserves of

For years, claims by North Korea of the possession of large amounts of crude oil were dismissed by the oil industry worldwide but recent surveys in which western organisations have been involved have confirmed that North Korea has much offshore oil. Four test wells in the West Sea have proved productive and reserves in the East Sea, also known as the Sea of Japan, have also been identified. While parts of the Sea of Japan are in North Korean waters, other parts are in Japanese waters and it remains to be seen whether Japan itself will attempt to increase its currently very low offshore hydrocarbon production, limited to **Honshu Island**, by undertaking exploration in this region.

In comparison, South Korea has no domestic oil reserves at all, and just one offshore natural gas field which came into operation in 2003. South Korea is consequently a major importer of hydrocarbons.

Northampton, PA

The location of steam generation to service a nearby pulp and paper mill using a **fluidised bed** with **culm** as the fuel. With heterogeneous fuels such as culm, laboratory measurements of the **calorific value** are not applicable because of the difficulty in obtaining a representative sample. The performance characteristics of the plant using the fuel therefore have to be used in what is known as the 'boiler-as-calorimeter' method. In the example below this is attempted for culm on the basis of information for the Northampton facility.

The fluidised bed facility at Northampton produces 277 MW of heat and uses 545 000 tonnes of culm per annum. Letting the calorific value be Q MJ kg^{-1},

$$277 \times 10^6 \, \mathrm{Js}^{-1} = \frac{545000 \times 10^3}{24 \times 3600 \times 365} \, \mathrm{kg}\,$$

$$\Rightarrow Q = 16.0 \, \mathrm{MJ \, kg}$$

This calculation has not accounted for losses which a full boiler-as-calorimeter treatment would incur using temperature measurements made at the plant.

Northeast heating oil reserve

A 2 million **bbl** reserve of distillate heating oil is kept for use during shortages in the populous north eastern states of the US, having been in place since 2001. It is divided between four terminals: 1 million bbl are at a terminal in Woodbridge, NJ, 0.75 million bbl at two separate terminals at New Haven, CT, and 0.25 million bbl at a terminal at Providence, RI. The reserve is seen as an extension of the **Strategic Petroleum Reserve** and was initially stocked by exchanging crude from the Reserve for distillate fuel oil. The Northeast heating oil reserve will supply the region with heating oil for up to ten days. President George W. Bush is currently discussing with advisors the setting up of a reserve of gasoline.

Northern California, oil and gas production in

Notwithstanding the fact that the drilling of the first oil well by **Drake** was on the eastern side of the US, it is often asserted that oil drilling at a commercial level was developed in California and that activity there goes back as far as the industry itself. It was in fact in 1865, just six years after the Drake well, that the first well in northern California was drilled.

In present-day northern California, there are a number of productive fields close to Sacramento including the Sacramento Delta field and the Union Island field, with respectively eight and five producing wells. The Rio Vista gas field, northeast of Sacramento, contains roughly one **bcm** of non-associated gas.

Northern Cascadia

A region 100 km from Vancouver Island, with reserves of **natural gas hydrates** which are currently being appraised. Wells have been drilled at five sites at water depths of 300 m. The current appraisal is the second of two off Vancouver Island: this area is seen as a possible site for an eventual gas production facility. There is similar activity off the Oregon coast.

Norway, entry of into coal-fired power generation

At the present time, power generation in Norway has been almost entirely hydroelectric. There are plans for a 400 MW coal-fired facility to commence electricity production in western Norway in 2011. Carbon dioxide from the coal combustion will be taken to depleted oil fields such as the **Sleipner field** for burial. The proximity of such sequestration sites to the coal-powered electricity facility is a considerable factoring the viability of the latter.

Norway, oil and gas reserves

Norway is the world's third-largest exporter of oil and gas (using only 5% of its own production) and the petroleum industry accounts for approximately one third of state income. Norway contributes 2 percent of the world's gas production and is also one of the 5 largest gas producing nations. Over the last 40 years Norway has become adept at operating in the challenging North Sea environment and has become a world leader in both technological and environmental issues.

In 2007, production from two large gas developments will begin. Snøhvit is a gas field located at a depth of 340m in the Hammerfest Basin and contains **condensate** and an underlying oil zone. Gas reserves are estimated at 160 **bcm** and after treatment the CO_2 extracted will be reinjected into the field. Ormen Lange is located at depths of between 800 and 1100 metres, off the coast of mid-Norway and contains 375 billion cubic metres of gas. Twenty-four wells are planned for Ormen Lange, in four seabed templates. The untreated well stream will be directed through two multiphase pipes to an onshore facility at Nyhavna, where the gas will be dried and compressed before being sent 1 200km to the United Kingdom, through the world's longest offshore gas export pipeline

NO_x

Formation of NO_x in flames is by three routes: **thermal NO_x, fuel NO_x,** and **prompt NO_x.** In **pulverised fuel** combustion thermal NO_x accounts for 5–25% of total NO_x formed, fuel NO_x accounts for 70–80% and prompt NO_x accounts for < 5%. NO_x reduction methods include the installation of **low NOx burners**, **selective catalytic reduction**, **selective non-catalytic reduction**, and **SNOx process**.

Nuclear power reactors, world distribution of

Over thirty countries have nuclear power generation, for example there are about 100 nuclear power reactors in the US and 14 in Canada. Other countries with major activity in this area include France, with 59 nuclear reactors, and Russia with 29. There are 9 in Spain and 19 in Germany. There are about 60 nuclear reactors in Japan yet only 3 in China. There are 11 in India, and some world opinion is that expansion of these is preferable to the growth of conventionally-generated power which the installation of the **Iran-India Pipeline** would enable. There are two nuclear power reactors in South Africa. Other countries with only one or two such facilities include Brazil (1), the Netherlands (1), Mexico (2) and Argentina (2). A nuclear power station might have more than one reactor; for example, in the UK there are 31 nuclear power reactors and 14 nuclear power stations.

The total annual production of electrical energy from nuclear reactors worldwide is of the order 2500 TWh ($\approx 10^{19}$ J).

O

Oahu, Hawaii

Site of a very productive **waste to energy plant**, where power at 150–200 MW is produced from locally generated waste and passed along to the grid.

Oak Creek, WI

The location of a power plant under construction by Wisconsin Energy Corporation of Milwaukee, expected to commence supply in 2009. Completion was delayed by an initially successful legal challenge in 2004 on the grounds that coal will be used as fuel. This was overturned in mid-2005. The capacity will be 1230 MW, and Wisconsin Public Power Inc. and Madison Gas and Electricity have each undertaken to buy 100 MW, thereby obtaining in effect an 8% holding in the enterprise.

Obaiyed field

Non-associated gas field in Egypt's western desert, producing since 1999 and operated by Shell. Shell, who have been active in Egypt since 1911, also operate the El Din field which yields 16 000 **bbl** per day of oil and large amounts of associated gas.

Obrigheim, Germany

Scene of a nuclear power plant in Germany which supplied power to the grid from 1969–2005 using **mixed oxide (MOX)** as fuel. There was exploratory work at Obrigheim into the use of mixed **thorium**/plutonium fuel while retaining **fuel rods** designed for **MOX**.

Ocean energy
A term used to refer to any form of energy that can be extracted from the ocean. This includes mechanical energy from currents, tides and waves as well as thermal energy from the difference between the cold deep waters and warm surface waters. No useful method for exploiting the salinity gradient in the ocean has yet been found.

Octane rating
The percentage of **iso-octane** in a mixture of iso-octane and n-heptane, which has the same knock characteristics as the gasoline fuel under consideration. 'Knock' refers to ignition ahead of the spark, leading to the impairment of performance and possibly extreme damage to a vehicle's engine. Although **iso-octane** is very resistant to knock, being capable of releasing reactive centres to eliminate the chemical intermediates responsible for knock, as n-heptane is a straight chain compound, it does not form reactive centres readily and is very poor at resisting knock. No gasoline offered for sale is as poor in terms of knock as n-heptane. Equally, very few are as good in terms of knock as iso-octane, and so a blend of the two forms a good basis for quantifying knock performance. Research octane number and motor octane number are determined at different rpm.

Oct-1-ene
Organic compound of formula CH_2=CH–C_6H_{13}, an example of an **alpha olefin**. It has the additional use of being a co-polymer with **ethylene** in the manufacture of some forms of polythene. **Sasol** are currently building a plant which, when fully operational, will produce 10^5 tonnes per year of Oct-1-ene.

Oil crisis
Formally defined as an increase in oil prices large enough to cause a worldwide recession or a significant reduction in global gross domestic product (GDP) below projected rates by 2–3%. There are two possible causes. A rapid expansion in global economy could result in a greater demand for oil or lack in spare production capacity such that demand would exceed supply. The second cause can result from a sudden disruption to oil supply owing to political events and decisions or

acts of terrorism. There have been oil crises in 1973 and 1979. Fears are that oil may reach $100 a barrel by 2008, which will trigger a global economic collapse.

Oil cut

A mixture of crude oil and **drilling fluid** generated in the drilling of an exploration well.

Oil from shale

Shale consists of a band of organic material called kerogen within a rock structure. The shale can be crushed and retorted whereupon crude shale oil is yielded by decomposition of the kerogen. The term shale oil means crude oil derived from shale in this way and this is usually hydrogenated before refining; there might also have been sulphur removal. Refining of this material gives fractions corresponding to those from crude oil and interchangeable with them in fuel utilisation. Shale usage significantly predates crude oil usage; there was use of shale-derived fuels in countries including Scotland and France by 1850.

A difficulty with production of oil from shale is the disposal of the rock after retorting (spent shale). That is why the truly enormous shale oil reserves of the world, which significantly exceed those of conventional crude oil, have only ever been used to a very limited degree. The viability of production from any particular shale deposit depends on the yield of crude shale oil per tonne of unprocessed shale. This obviously depends in some degree on the efficiency of processing as well as on the deposit, and figures in the range 50–150 kg oil per tonne of shale are typical for those deposits which have been worked. Note that this does not translate to 5–15% of the weight of the shale, as the oil will almost certainly have been hydrogenated after retorting.

The US has massive reserves of shale, enough to yield on processing about 500 billion **bbl** of shale-derived crude oil. It is widely distributed, there being particularly abundant deposits in states including Colorado and Utah. (It is said that if one approximates the 48 mainland states of the US to a rectangle and draws a diagonal from northeast to southwest, that represents a continuous deposit of shale of varying weight ratios of kerogen to rock.) Although there has been shale oil production of local importance in the US over the decades it never built up into a major industry and at present there is no significant production of oil from shale.

The table below lists details of shale oil production from five selected countries.

Country	Crude oil production from shale $(kTyr^{-1})$	Comments
Australia	5	Some export to Asia
Brazil	195	Production for the domestic market since 1935
Estonia	151	Significant exports as well as domestic use in applications including power generation
Germany	500	One deposit only at Dotternhausen yielding this relatively large quantity. Spent shale diverted to cement manufacture
Israel	450	Low yields of oil, which is also high in sulphur. Nevertheless, successful power generation in the Negev region of the country with locally won shale-derived fuels

Other countries with activity in the area of shale-derived fuels include China, where a new retorting facility began production in 1992. **Carbon black** is a by-product of the production of fuels at the plant, which is called the Fushun Oil Shale Retorting Plant. There had in fact been shale oil processing at Fushun previously with less advanced plant; the city is also a major coal centre. There has been much exploration, with promising results, for shale in Morocco where there is the proven potential to raise over 3 billion **bbl** of shale-derived crude oil. There has not as yet been any actual production, possibly because the deposits are in isolated areas lacking not only infrastructure but also water which is needed for the processing. The moderate reserves of shale in Thailand, to the north of the country, have not been developed and this again is probably due to their isolation from centres of population.

There have been failures in the past at attempts to produce oil from shale in quantities comparable to conventional liquid hydrocarbons, one of which ended on **Black Sunday** in 1982. This means that the winning of the confidence of investors and participants in any such enterprise in the future will be very important.

Oil India Ltd.

This company operates in northern India and its primary asset is a pipeline of length 1157 km working in the east–west direction. The pipeline, which has been in existence since 1962, delivers oil along its length to five refineries, the eastern-most refinery being **Digboi**. At the present time about half of the oil carried along this pipeline is from wells previously deemed depleted, **enhanced oil recovery** techniques having been used. The pipeline along its route comes very close to the northernmost tip of Bangladesh.

Oil reserves, rate of discovery of

All of the major oil companies of the world include exploration among their activities and at present their combined efforts add up to about 10 billion **bbl** per year, a figure which is for conventional crude plus **natural gas liquids**. The figure of 80 million bbl per day for production of oil in the early years of the 21st Century given in a previous entry converts to just under 30 billion bbl per year. The inequality is the basis of some discussions of the Oil Peak. However, this peak does not signify that amounts of oil discovered in new fields are falling well behind amounts removed from existing fields, but that 50% of whatever proven resources there are have been used up. Here, identification with **Hubbert Peak** is possible, but the significance is not its own occurrence—that is inevitable, even though estimates of its date vary widely—but the fact that according to a hypothesis it is the stage at which world oil production will start to decline.

It is interesting to examine discovery figures over the decades, first noting that a plot of such discoveries across the years has many spikes in it, reflecting major finds. The steepest spike is in about 1950, when exploration at the Gulf Coast and also in the Middle East was intense. Discovery was then a staggering 100 billion bbl per year. This figure was approached once more in the 70s, contributed to by the North Sea.

Oil Rivers Protectorate

Name applying between 1891 and 1900 to a region of southern Nigeria, part of the then British Empire. Counterintuitively, given that Nigeria is now such a major oil producer, the name derives not from crude oil but from palm oil which was the major export from the region at that time. Let it be noted that over 100 years later palm oil is very marketable as a fuel.

Oil sands

Deposits of bitumen that will not flow unless heated or diluted with lighter hydrocarbons. Canada ranks second largest in terms of global proven crude oil reserves (15% of world reserves) after Saudi Arabia.

The majority of these reserves are found in Alberta's oil sands (over 174 billion barrels); this could satisfy the global demand for petroleum for over a century. Bitumen makes up about 10–12% of the actual oil sands found in Alberta. The remainder is 80–85% mineral matter and 4–6% water.

Oil and gas pipelines, integrity of

Half of the oil and gas pipelines in the USA are older than 40 years old and some are older than 70 years old. There have been fatal accidents in recent years, including the accident at **Carlsbad, NM**, which have all originated from pipeline leakage. The remedial approach currently being taken, with regard to increased demands for natural gas in the US, is the use of quantitative risk analysis to provide a means of managing old pipelines. Such quantities as the design stress of the current pipeline material can be estimated as can the probability of failure of any feature or extension not original to the pipeline e.g. a weld repair. Any obviously necessary work such as the replacement of a section containing a crack can be carried out ahead of the risk analysis.

The bottom line of the quantitative risk analysis is the frequency of a fatal accident due to pipeline failure. This can be compared to the corresponding figure for a newer—perhaps unused—pipeline of the same specifications. The age of a pipeline will not feature in such an analysis, meaning that the comparison of an old pipeline with a new one is based solely on engineering criteria. One would expect the risk to be lower with the new pipeline, but the risk due to the older pipeline need not be so high as to make its use, for an agreed time, until the next risk analysis unacceptable. By this soundly based approach, many elderly pipelines are continuing in safe use.

Olefins

Unsaturated, straight or branched chain hydrocarbons with the general formula C_nH_{2n}, otherwise known as *alkenes*. They contain one or more double bonds obtained from petroleum fractions by cracking at high temperatures. The most important olefins and diolefins used to manufacture petrochemicals are ethylene, propylene, butylenes and butadiene ($CH_2 = CH - CH = CH_2$).

Once-through

A heating process where water is pumped from the source through a heat exchanger and then discharged. There is no recirculation of the water as in the steam cycle in the coal-fired power station.

OPEC
See **Organisation of Petroleum Exporting Countries**

OPEC and non-OPEC pricing competition
The *raison d'etre* of **OPEC** was solidarity of member countries, in particular assertion of rights to set their own prices for their oil. There are times when the non-OPEC nations have to respond to the need to counter the decisions of OPEC. If the OPEC prices are transiently high, importing countries will appeal to non-OPEC producers (often including Russia) to step up their production. That will mean that importing countries take less OPEC oil and more non-OPEC, which puts some pressure on OPEC to curb its oil price increases. This 'push-pull' between OPEC and non-OPEC countries is a feature of oil marketing.

OPEC, possible extension to the membership of
At the time of going to press, Angola and Sudan have both been offered membership. If either or both accept, it will be the first admission of a new member since Nigeria in 1971.

Organisation of Petroleum Exporting Countries (OPEC)
The hydrocarbon-producing countries of the Middle East entered the oil industry much later than, for example, the US and the former USSR. By 1962 when the United Arab Emirates made its first export of oil the motorcar had proliferated and the demand for gasoline was high. Similarly, the petrochemical industry was well established, cracking technologies having been developed 30 years earlier. Polymers had also to a large extent already replaced wood and natural fibres in the manufacture of household fittings and clothing etc.

The discovery of gigantic amounts of oil in the Middle East and, at about the same time, in certain developing countries in other parts of the world, required a safeguard that prices paid for oil from such countries would be the same as those for oil from those with established firms. It was also necessary to ensure that expatriate firms investing in the new oil-producing countries were protected and assured of fair return on plant installed and services provided. With these objectives in mind, **OPEC** was formed in September 1960 at a meeting in Baghdad. The five original members were Iran, Iraq, Kuwait, Saudi Arabia and Venezuela. Eight other countries later joined: Qatar, Indonesia, Libya, the UAE, Algeria, Nigeria, Ecuador (withdrew 1992) and Gabon (withdrew 1994).

The world at large became aware of OPEC and its potential for impact on world affairs in the early 1970s. By that time the US, which less than 25 years previously had been self-sufficient in oil, was importing 35% of its crude oil, mainly from OPEC countries. There was military action against Israel by Arab groups, and ultimate victory for Israel was secured by US intervention. OPEC responded by decreeing that no oil would be supplied to the US and that price rises of 70% would apply to America's allies in Western Europe. This continued until March 1974, the disruption to life in the affected nations being enormous. The aggregate production of crude oil by the OPEC nations (December 2006) is 33.5 million **bbl** per day (exceeding the official quota of 28 million bbl per day), about 35% of the world total.

Oriente
Region east of the Andes in Peru and the site of one of the major oil fields of that country, accounting for about two-thirds of its production of crude. Remaining reserves of oil in Peru are fairly small, having been estimated as 350 million **bbl**.

Orimulsion®
Trade name for an emulsion comprising **bitumen** ($\approx$70%) with water ($\approx$30%). There is also an emulsifier. It was made (production was discontinued in December 2006) in Venezuela from locally produced bitumen and used as a substitute for fuel oil in applications including steam raising.

Ormen Lange
A gas field under development in the Norwegian sector of the North Sea. The production platform will only be about 100 km from the coast, and a terminal at Nyhamna in northwest Norway will receive the gas. Companies with an interest include Shell, BP and Exxon Mobil.

Orsat apparatus
Classical wet chemistry method of analysis of post-combustion gas. Such gas is passed successively through solutions of potassium hydroxide, pyrogallol and cuprous chloride, totally removing the carbon dioxide, oxygen and **carbon monoxide** respectively from the post-combustion gas. The method has the advantage that the entire assembly of reagents in tubes is mounted on an easily portable frame.

Oscillating water column (OWC)
Device which provides air under pressure to turn turbines in the gen-

eration of power from **marine wave energy**. Seawater enters the OWC and displaces air which itself turns the turbine blades.

Outage
A temporary loss or suspension, intended or otherwise, of electric power.

Outer Brewster Island
Ten miles out to sea from Boston, currently occupied by redundant military plant, Outer Brewster Island is proposed as the site of a huge storage facility for **liquefied natural gas (LNG)** which could be transferred ashore along an existing **pipeline**. An argument in favour of the Outer Brewster Island plan is that an LNG vessel actually in Boston Harbour could be the target for a terrorist attack as there are residential premises very close to the Harbour.

Overburden
The rock or soil above a useful mineral (or coal) deposit that must be removed before commercial mining can take place. The amount of overburden that must be removed in relation to the amount of coal that will be obtained from the bed below is known as the overburden ratio.

OWC
See **oscillating water column**

Oxidative coupling
Conversion of a C_1 species such as methane to a C_2 species such as ethanol by reaction with oxygen under conditions such that full oxidation is precluded. Such a reaction might be represented by:

$$2CH_4 + O_2 \rightarrow C_2H_5OH + H_2O$$

Oxygenate
An oxygen-containing chemical that is added to fuel to help it burn more efficiently and hence produce less pollution. The two most common oxygenates are ethanol and **methyl tertiary-butyl ether**. The term oxygenation is also used to describe the process of combining or treating a substance with oxygen.

P

Pailin field

Non-associated gas field in the Gulf of Thailand, like the **Trat field** operated by the Uncol Corporation with certain other bodies, including Mitsui Oil, having an interest. As with the Trat field the gas is for local use largely in power generation. As at other places including **Freeport, TX**, some of the methane produced is diverted to petrochemical manufacture. It is worth noting that at two very disparate places methane is being used in petrochemical manufacture reflecting a growth of such use. Amerada Hess have a holding in the field.

Pakri wind farm

Located on the coast of Estonia, supplying an impressive 1% of that country's power needs. The town of Paldiski where the wind farm is situated was once the westernmost limit of the eastern bloc and as such was under very close surveillance by the USSR who, in the cold war era, would have 16 000 soldiers stationed there at any one time.

Palladium chloride

Reagent used to detect and semi-quantitatively measure **carbon monoxide** in the atmosphere. In modern applications the compound is supplied dissolved in a gel in a small sealed unit. On breakage of the seal and the drawing of air into contact with the gel, carbon monoxide causes deposition of metallic palladium, easily visible. When in an industrial setting there has been leakage of process gas which might have raised the CO level there might not be time to engage an instrumental method of detecting the compound, such as an infra-red device. Techniques involving palladium chloride or **hoolamite** are valuable in such contingencies.

Palm oil

Currently being widely produced internationally as a component of **biodiesel**. The world's major producer at present is Malaysia where ≈ 9 Mt of the material are being produced each year and a 10% growth is expected by 2010. In peninsular Malaysia, where land is at a premium, new palm plantations can only be established where rubber or cocoa were previously grown. New palm plantations are therefore in Sarawak and Sabah, where returns of 4 tonnes of palm oil per hectare per year are expected. Fluctuations in the demand for palm oil are due largely to the competition from **soybean** oil as a basis for biodiesel. The world's second largest producer of palm oil is Indonesia.

Panama Canal, movement of crude oil through

Around 500 000 **bbl** of crude oil per day is moved through the Panama Canal, almost two thirds of it in the Atlantic–Pacific direction.

Panamax

The largest size of fully-laden oil tanker that can safely navigate the Panama Canal due to the limitations of the locks which are 33.53 m wide, 320 m long and 25.9 m deep. The typical displacement of a Panamax cargo ship would be around 65 000 tonnes.

Panhandle

One of the most productive oil and gas reserves in the world, this vast Texan field began production in 1918. In 1994, Panhandle field, the largest-volume gas field in the US, reported annual production of 165 664 617 000 ft^3 of gas from 4499 producing wells and an annual oil production of 5 023 878 **bbl**.

Paper waste, fuel use of

There is fuel use of paper in countries including the Netherlands, where a major paper manufacturer produces fuel pellets from the waste generated, sold as ROFIRE®. A quantity of 30 000 tonnes per year of paper sludge waste is so treated leading, after drying and beneficiation, to 16 000 tonnes of ROFIRE® pellets with a **calorific value** of 23.7 MJ kg⁻¹, significantly higher than that of pellets made from **municipal solid waste**. A company in Slough, England, is marketing **briquettes** made from paper and cardboard.

Papua New Guinea, hydrocarbon reserves of

This nation, which gained independence from Australia in 1975, has known reserves of oil of 400 million **bbl** and of natural gas 0.6 million m³. A pipeline for export of natural gas to Queensland, Australia, is planned. There is a refinery operated by **InterOil** which processes 30 000 bbl per day of crude oil: about half of its products are exported.

Para xylene

A component of **benzene, toluene, xylenes (BTX)** and a starting material for the production of purified terephthalic acid (PTA) for polymer manufacture. Para xylene can be made from the **naphtha** fraction of crude oil and this does in fact take place at a number of refineries including that at **Sriracha** where 350 000 tonnes of the chemical are produced annually.

Paraffins

Saturated, straight or branched chain **hydrocarbons** with the general formula C_nH_{2n+2}, otherwise known as alkanes. Obtainable from petroleum, the simplest is methane CH_4.

Paratus power station

Power station in Namibia comprising a number of **gensets** with a capacity of 20 MW, brought into use on a standby basis for the coastal areas of the country. Namibia has 240 MW of **hydroelectricity** and 120 MW of thermally generated electrical power. Though a net importer of electricity, Namibia also exports some to Angola and to Botswana. Namibia also supplies top-up power to South Africa during the winter.

Patent fuel

A fuel manufactured by compressing small coal particles with a binding agent into a solid briquette.

Paute

Site of **hydroelectricity** at 1.1 GW level in Ecuador. Its operator is the state-owned Hidropaute. Hidropaute are currently building a supplementary 180 MW facility a few miles up river from the Paute plant, which will enable supply to be boosted as necessary. Over half of the electricity for this country (population 13.5 million) comes from the Paute installation.

Peabody Energy

Founded in 1883 in Chicago, now the largest private coal producer in the world. The amount produced in 2004 was in excess of 200 Mt. The company is active in several parts of the US and also in Queensland, Australia and Venezuela. The company is, at present the time, negotiating to commence activity in China.

Peak load

The maximum amount of power given out or taken in by a machine or power distribution system over a certain period of time. Peak load plant is normally only operated during periods of highest demand and usually consists of older plant. For example, a coal-fired power station may be brought online for 3 pm and taken offline at 9 pm between October and March, ensuring power demands are met in the winter. Necessary maintenance would be carried out during the summer months.

Peanut shells, fuel use of

Peanut shells have been used to make **producer gas** and there has been very limited activity in pyrolysing them to make a richer gas than producer gas, with **char** and tars/oils as by-products.

At the present time, the country with the greatest involvement in fuel use of peanut shells is Senegal where there have been feasibility studies into steam raising, steam so produced undergoing a **Rankine cycle** to make electricity. It has been shown that 22 MW of electricity could be provided for Senegal in this way. This could be increased by supplementing the peanut shell fuel with the **bagasse** produced in that country.

Peat

The precursor to **lignite** in the **coalification** sequence, formed by the conversion of cellulose in vegetation material (from which **coal** is ultimately formed) to humic acids. Peat is used as a fuel for homes in many parts of the world, notably the former Soviet Union and China. Peat is always very wet when first cut from the ground. Air-drying yields a solid fuel with a calorific value of about 15 MJ kg^{-1}. It can be further improved by forming into **briquettes**.

In the early 21st century, a number of countries are still involved with the utilisation of peat as fuel on a large scale. A case in point is Belarus, where most of the peat is made into briquettes for heating applications. Production of such briquettes in Belarus is 1.7 Mt yr[-1], and there is some export to neighbouring countries. Estonia also produces peat for manufacture of briquettes, but on a scale about an order of magnitude lower than Belarus. The Sveg region of central Sweden is also a scene of peat production and manufacture of briquettes. Peat can be used in large-scale electricity generation using combustion plant initially designed for low-rank coals. There is currently power generation from peat as **pulverised fuel** in Ireland and also in Finland where peat is used in **combined heat and power** applications as it is in countries including Latvia and Sweden.

Pecan nut

Tree nut native to the US. Pecan shells find fuel application similarly to other biomass fuels such as **peanut shells** or **rice husks**, but wood from the Pecan tree has a more specialised use in food smoking, being particularly suitable for poultry, beef, pork and cheese.

Pecos County, West Texas

Scene of electricity generation with **wind turbines,** there being 125 originally manufactured in Denmark by **Vestas**, generating altogether 150 MW. Connection to the transmission system of the local electricity facility is in place. The turbines are sited on land used for grazing with some oil and gas beneath it. None of the current uses of the land will be affected by the presence of the turbines.

Pelletised wood

Solid fuel commercially available in a number of countries including Sweden, where 800 000 tonnes of such fuel are produced annually with applications including steam raising. There is also significant usage in the UK. As a result of its compression in manufacture, pelletised wood has a lower moisture content than fuel wood which has only been naturally seasoned, and consequently pelletised wood has a **calorific value** of > 20 MJ kg[-1]. Usually no binder is needed to pelletise the starting wood waste material as the lignin in the wood fulfils this role. Pellets are typically cylindrical with a length of 2 cm and a diameter of 6 mm. Unlike some binderless coal **briquettes**, pelletised wood fuel does not degrade or disintegrate on storage.

Pemex

Mexican petroleum company, formed in 1938 and government owned. Pemex produces 3.4 million barrels of oil per day, two-thirds of it offshore, and 90 % of the oil is exported to the US. Accordingly, other petroleum companies have arrangements with Pemex where they purchase the oil themselves and undertake downstream activity. For example, Exxon-Mobil buy 435 000 barrels per day of crude oil from Pemex and process it at their Gulf Coast refineries. Shell has a similar arrangement with Pemex, whereby Mexican oil is taken to the Shell refinery at Deer Park near Houston, which has been in operation since 1929. This refinery is sometimes referred to as the Shell-Pemex refinery.

Penészlek field

A natural gas field in eastern Hungary having produced 0.1 **bcm** gas between 1983 and 1989 from six wells. New wells are now being appraised by London-based Ascent Resources, which also has exploration activities in Italy, Switzerland, the Netherlands and Romania.

Percussion drilling

Drilling at an oil well by pounding with a **bit** as opposed to rotating the bit. It was a common technique up to about 1920 but has now been totally replaced by rotary drilling.

Persian Gulf, shipping times of crude oil from

From the Persian Gulf to the Gulf States of the US takes about 40 days while to Japan it takes around 20 days.

Perth Amboy NJ

Site of an acetylene manufacturing plant where in January 2005 there was an explosion in which three people were killed and one other seriously injured. Leaked acetylene drifted to a shed external to the plant where the surface of a space heater was believed to have provided an ignition source.

Perusahaan Listrik Negara (PLN)

State operated supplier of electricity for a large part of Indonesia. There is some **geothermal electricity** in Indonesia and this accounts for about 5% of the total. Other sources managed by PLN are **hydroelectricity** (6%), the balance being thermal generation using fuel oil or natural gas. PLN supplies in total just under 22 GW.

Petrochemical
The term for the range of chemicals or other materials that can be derived directly or indirectly from natural gas or petroleum.

Petrol station pollution
Term applied to the release of fuel vapour from filling stations. Such releases are believed to amount to about 16 000 tonne per year in the UK. Devices to capture the vapour and recover it are routinely available and from October 2007 all filling stations in the UK selling more than 3.5 millions litres per year of petrol will be required to use such a device.

Petroleum coke
The solid material remaining when the heaviest (i.e. highest molecular weight) part of crude oil is pyrolysed. Oil gas tar is a by-product. One common use of petroleum coke is as carbon electrodes in industrial electrolysis processes.

Petroleum, origin of
The word here is used in the generic sense suggested by its etymology, which is simply 'rock oil', and is therefore considered to include crude oil, **heavy crude** and natural gas. The origin is believed to be very basic organisms at the surface of the sea similar to what we now know as plankton. Once such an organism had died, it would have descended to the bottom of the sea. Deposition of organic matter at the seabed in this way would be followed by burial under **sedimentary rock** and over time an organic deposit buried under the sea resulted. The action on this by anaerobic bacteria, on a time scale similar to that of **coalification**, led to the formation of petroleum.

There are however two major differences from coalification. One is that pressure and temperature conditions are less important in petroleum formation than in coalification. The second is that coal, being a solid deposit, remains at the site where the initial plant deposition was (apart from drift coal) whereas petroleum, being a fluid, migrates through the rock structure. Its eventual place of discovery in some underground reservoir will be quite distant from the site of the initial organic deposition.

Petrolimex
Vietnamese company, with its HQ in Hanoi, which is concerned with the import and distribution of refined petroleum products into Vietnam. There is the anomaly that although Vietnam is a major oil pro-

ducer it lacks any domestic refining capability, therefore refined products have to be imported and Petrolimex perform about two-thirds of such importation. There are plans to build a **grass roots refinery** for Vietnamese crude oil at **Dung Quat**. Planning has been delayed and lead to many frustrations; the import of refined material is likely to be necessary for several more years.

Among the assets of Petrolimex are six large tankers for long-distance use and many other smaller vessels. The company operates just over 1000 retail gasoline outlets in Vietnam and has a fleet of road tankers for deliveries to them. Some of the fuels it brings to Vietnam it subsequently re-exports to Laos, Cambodia and southern China. It owns many liquid hydrocarbon storage vessels manufactured by the company itself, 500 km of pipeline and has five major storage facilities for **liquefied natural gas**. It has also diversified into business quite unrelated to oil and gas, such as insurance.

Petronet

Indian company, half-owned by India's four state-controlled oil companies. Gaz de France have a significant interest and, the company having entered the stock market in 2004, the remainder is held by shareholders. The company's activities are all in distribution: it has no involvement in production or refining. A major part of its remit is pipeline installation and maintenance, also the import of **liquefied natural gas** from Middle East sources. It also has a major role in the proposed **Dahej terminal**.

Philippines, oil and gas reserves

This Asian country has a number of offshore oil fields in the South China Sea. The most productive and best known is the field operated by the Australian concern Nido, which began production in 1979. By March 2005 it had produced 17 million **bbl** of oil. In the same country, Nido will be developing the oil fields Palawan Basin, where exploration wells are currently being drilled. The Camago-Malampaya field, which is in deep water, is the country's first commercially viable gas field and Occidental (who discovered the field) and Shell are both active there. Texaco have a significant share. The gas is already being brought ashore for use in thermal power stations at 1500 MW level.

Phosphoric acid fuel cell

Phosphoric acid fuel cells use an electrolyte that conducts hydrogen ions (H^+) from the anode to the cathode. The electrolyte is composed of liquid phosphoric acid within a silicon carbide matrix material. Phosphoric acid fuel cells operate at about 150–205°C and a pressure of about 15 psig (1 barg). Each cell can produce up to about 1.1 VDC.

The advantages of phosphoric acid fuel cells are that they:

• are tolerant of carbon dioxide (up to 30%). As a result, phosphoric acid fuel cells can use unscrubbed air as oxidant, and reformate as fuel

• operate at low temperature, but at higher temperatures than other low-temperature fuel cells. Thus, they produce higher-grade waste heat that can potentially be used in co-generation applications

• have stable electrolyte characteristics with low volatility even at operating temperatures as high as 200°C.

The disadvantages are that they:

• can tolerate only about 2% carbon monoxide

• can tolerate only about 50 ppm of total sulphur compounds

• use a corrosive liquid electrolyte at moderate temperatures, resulting in material corrosion problems

• have a liquid electrolyte, introducing liquid handling problems. The electrolyte slowly evaporates over time

• allow product water to enter and dilute the electrolyte

• are large and heavy

• cannot auto-reform hydrocarbon fuels

• have to be warmed up before they are operated or be continuously maintained at their operating temperature.

Phosphoric acid fuel cells react hydrogen with oxygen. The reaction at the anode is:

$$H_2 \rightarrow 2H^+ + 2e^-$$

The reaction at the cathode is:

$$\tfrac{1}{2}O_2 + 2e^- + 2H^+ \rightarrow H_2O$$

The H^+ ion is drawn through the electrolyte from the anode to the cathode by the reactive attraction of hydrogen to oxygen, while electrons are forced through an external circuit. Combining the anode and cathode reactions, the overall cell reaction is:

$$H_2 + \tfrac{1}{2}O_2 \rightarrow H_2O$$

Therefore, the fuel cell produces water that accumulates at the cathode. This product water must be continually removed to facilitate further reaction.

Photovoltaic cell

A device which uses solar radiation to generate electrical power. Applications range in scale from a watch or a calculator to a power station.

Solar flux is about 1400 W m^{-2} of the receiving surface (that is, the Earth's surface where solar irradiation is occurring) at high sun, less at other times of day. In a photovoltaic cell such irradiance on to a semiconductor produces a current. The semiconductor will often consist of pure crystalline silicon oxide into which a few deliberate impurities, such as elemental boron or elemental gallium, have been introduced at ppm level. Such a medium responds to solar radiation by movement of electrons. Conversion of solar energy to electrical will be 10–15% efficient, the balance simply being manifest as heat.

Clearly the area receiving solar flux is an important quantity in photovoltaic cell performance. Often a cell of standard area is used. The number of cells required to form a composite cell for a particular application is determined via the calculation below.

A photovoltaic cell is to be manufactured from standardised silicon semi-conductor units, each with an area, for irradiance purposes, of 10 cm × 10 cm. If the solar flux received is 1000 W m^{-2} and the power required from the composite cell is 3 kW, how many of the individual cells are needed if the efficiency is 12%?

Letting the number of 10 cm × 10 cm cells required be n and the subscripts 's' to denote solar power and 'p' electrical power:

$$1000 \, W_s \, m^{-2} \times 0.12 \times (n \times 0.1 \times 0.1) m^2 = 3000 \, W_p$$

$$\Rightarrow n = 2500$$

so 2500 of the small cells would be needed to deliver 3 kW. This is a fairly large-scale example: the power requirement of a watch or a calculator would of course be many orders of magnitude lower than that in this calculation.

Photovoltaic cells are a growth industry. Major projects are underway in countries including the US, Japan and Germany, involving the powering of buildings and facilities from photovoltaic cells. The Co-operative Insurance Society (CIS) Building in Manchester has over 7000 photovoltaic panels which generate over 180 000 kWh per year. There is also significant development work on photovoltaic cells in Puerto Rico where there is presently heavy dependence on imported fuels for power generation. The most ambitious undertaking in power generation from photovoltaic cells is that by **Socal Energy**.

Phu Horm
Thai non-associated gas field in the northeast of the country having recently commenced production at a rate of 1–2 million m³ per day, expected to double as new wells come into operation. All of the gas is to be used for the generation of electricity.

Piapvav, India
Site of a major **liquefied petroleum gas (LPG)** terminal with a capacity of 90 000 Mt per annum, owned and operated by **Shell** under the name Shell Gas (LPG) India Pvt Ltd. Shell's LPG activities in India had to be transferred to government oil companies for a period as a result of the nationalisation of the industry. Subsequent legislation enables parallel private operators to exist alongside state-owned operators, the basis of the supply of LPG by Shell at the present time. Shell has exclusive supply contracts for LPG with many of the major hotel chains in India. LPG is also used in India in the textiles, food, metal and ceramics industries. Piapvav is in the Gujarat region of India, which has a coast with the Arabian Sea. **Hazira**, site of the new **liquefied natural gas** terminal, is also in the Gujarat region.

Pico Canyon
Located within the Santa Clarita Valley in California and the scene of the first commercially viable oil in the western US in 1876. Production of oil continued there until 1990, by which time the well was owned and operated by Chevron.

Pierce County, ND
Noted over several generations for its abundance of lignite and its activity in the utilisation of lignite, North Dakota is also establishing itself as a centre for wind energy. To the existing wind power facilities there, which include one in Lamoure county with 41 turbines in operation, is being added another in Pierce County which will have 100 turbines with a combined capacity of 150 MW.

Pipeline magnet
Device by means of which any ferrous material such as rust, having transferred from the inside surface of a pipeline to the oil which it conveys, can be removed. The body of the device is made from stain-

less steel within which are non-ferrous magnets composed of elements in the lanthanide series. The oil in the pipe is filtered through the pipeline magnet and ferrous debris removed.

Pipeline repair clamp
A means of permanently repairing a damaged sub-sea pipeline. Its fitting requires a remotely operated vehicle (ROV), and power for the operation is provided hydraulically by the use of seawater.

Piper Alpha
Oil and gas production platform in the North Sea, operated by Occidental. In July 1988 it was destroyed in a fire and 167 lives were lost.

Piracy, oil tankers
A recent such incident (June 2006) involved an oil tanker operated by a Fujairah company (Akron Trade and Transport) with a crew of 19 Filipinos. The pirate attack occurred while the vessel was close to Mogadishu, Somalia. According to Pottengal Mukundan, director of the International Maritime Bureau, Somalia is "probably the highest risk area for piracy", with 40 attacks off the country's coast since March 15.

Plamuk field
Offshore oil field in Thai waters. Initial (2001) production was 2500 **bbl** per day, which increased to 18 000 bbl per day. The Thai offshore industry is the scene of an emerging technology for carbon dioxide removal from natural gas by a membrane system.

Plastic waste, fuel use
Plastics such as polythene, polystyrene and polypropylene have high **calorific values** but the use of plastic waste as a fuel is made difficult by the fact that such substances, if burnt as large pieces, form a melt thereby excluding oxygen and causing combustion to be highly incomplete. There is however some interest in co-combustion of coal and plastic waste as **pulverised fuel (pf)**. This of course requires milling of the plastic. Blends of plastic waste with a **bituminous coal** such as pf have been used on a trial basis in power generation. The range of composition of such blends has been from 5–20% of milled plastic.

Plastic waste, transport fuels from
This is a reality in Japan and shortly to begin in the UK, the technology having been developed in Japan and furthered in Australia. The plastic

waste is first pyrolysed, and in a patented catalytic process the pyrolysate reacts to form a liquid product interchangeable with diesel. The target in Britain is an annual 135 million litres of fuel from 136 000 tonne of waste plastic. A mass balance on these figures is shown below.

Let the density of the liquid product be ≈ 750 kg m^{-3}:

$$135 \times 10^6 \text{ litres} \equiv 135 \times 10^3 \text{m}^3 \equiv 1.0 \times 10^8 \text{ kg}$$
$$\equiv 1.0 \times 10^5 \text{ tonne}$$

Initial amount of plastic 136 000 tonne (1.36×10^5 tonne)

Extent of conversion = $(1.0/1.36) \times 100 = 74\%$

The remainder will be a solid or (more probably) semi-solid residue. A limit on the extent of conversion is imposed by the hydrogen-to-carbon ratio of the starting plastic material.

The method can use unsorted plastics, and the fuel resulting is lower is sulphur than mineral diesel fuels usually are.

Platelets™

Name given to a new technology for locating and sealing leaks in pipe-lines, the terminology obviously being derived from conceptual similarities with the action of platelets in the clotting of blood.

Small particles symbolically referred to as platelets are inserted into a pipeline and are carried by the fluid which the pipeline is bearing. As the fluid reaches the site of the leak, the platelets are diverted from the bulk flow and remain stationary at the part of the pipe's internal wall where the leak is. The leak position is then known and a repair can be affected, or the platelets themselves will seal the leak for a period sufficient for emergency measures such as pipe drainage.

Platform

An offshore structure from which oil and gas wells are drilled and exploited.

Pluto field

Newly discovered (April 2005) gas field off northwest Australia, in deep water. Appraisal wells are being drilled and it is intended that once production begins the gas will be made into **liquefied natural gas (LNG)** for export. The target figure is 5–7 Mt yr^{-1} of LNG.

PM$_{10}$

Airborne combustion-derived particles of diameter 10 μm or less, of which motor vehicles are a major source. Such particles lodge in the lungs when inhaled and can provide a route for entry to the body of **polyaromatic hydrocarbons.**

Point Thomson, AK

The location of a rich oil field, in close proximity to the **Arctic National Wildlife Refuge,** as yet undeveloped. Since it is not actually in the reserve there are no legal restrictions to drilling and the group of companies with drilling rights, including Exxon and Chevron, have come under criticism from the government of Alaska. The field is believed to contain 2 **bcm** of natural gas and several hundreds of millions of **bbl** of crude oil.

Polaniec power station, Poland

The site of the generation of electricity from coal and **forest thinnings** as a composite fuel. The national electricity company Electrabel have a total nationwide capacity of > 28 GW. The Polaneic installation will help significantly in attainment of the target of 12% of the nation's power from renewable sources.

Pollutant

A pollution-causing human-produced substance present in the environment. For example: **dioxin, NOx, soot, sulphur dioxide.** Pollution is the release of harmful contaminants to the environment. Although pollution can occur naturally, from volcanic eruptions for example, it is normally considered to be a negative effect of human activities. For example, car exhaust fumes, chemical spills, contamination of water supplies etc.

Polyaromatic hydrocarbons (PAHs)

Organic compounds having a structural framework of multiple benzene rings. They are synthesised in a flame by a mechanism involving fuel **pyrolysis** and are the precursor to smoke. PAHs are carcinogenic.

Polychlorinated biphenyls (PCBs)

A group of toxic, chlorinated aromatic **hydrocarbons** of organic compounds with 1 to 10 chlorine atoms attached to a biphenyl with chemical formula $C_{12}H_{10-n}Cl_n$. PCBs were once used in batteries, fire retardants, lubricants and paints, among other examples. They are no longer used due to their effects on health (e.g. cancer, birth defects and skin diseases) and long environmental life. Examples of structures include:

Polymer electrolyte membrane (PEM) fuel cell

Also known as proton exchange membrane or solid polymer fuel cells. This type of fuel cell uses an electrolyte that conducts hydrogen ions (H^+) from the anode to the cathode. The electrolyte is composed of a solid polymer film that consists of a form of acidified Teflon®. PEM fuel cells typically operate at 70–90°C and a pressure of 15–30 psig (1–2 barg). Each cell can produce up to about 1.1 VDC.

The advantages of PEM fuel cells are that they:

• are tolerant of carbon dioxide. As a result, PEM fuel cells can use unscrubbed air as oxidant, and reformate as fuel

• operate at low temperatures. This simplifies materials issues, allows a quick start-up and increases safety.

• use a solid, dry electrolyte. This eliminates liquid handling, electrolyte migration and electrolyte replenishment problems

• use a non-corrosive electrolyte. Pure water operation minimises corrosion problems and improves safety

• have high voltage, current and power density

• operate at low pressure which increases safety

• have good tolerance to differential reactant gas pressures

• are compact and rugged

• have relatively simple mechanical design

• use stable materials of construction.

The disadvantages are that they:

• can tolerate only about 50 ppm **carbon monoxide**

• can tolerate only a few ppm of total sulphur compounds

• require reactant gas humidification, which is energy intensive and increases the complexity of the system. The use of water to humidify the gases limits the operating temperature of the fuel cell to less than the boiling point of water and therefore decreases the potential for co-generation applications

- use an expensive platinum catalyst
- use an expensive membrane that is difficult to work with.

PEM fuel cells react hydrogen with oxygen. The reaction at the anode is:

$$H_2 \rightarrow 2H^+ + 2e^-$$

The reaction at the cathode is:

$$\tfrac{1}{2}O_2 + 2e^- + 2H^+ \rightarrow H_2O$$

The H^+ ion is drawn through the electrolyte from the anode to the cathode by the reactive attraction of hydrogen to oxygen, while electrons are forced through an external circuit. Combining the anode and cathode reactions, the overall cell reaction is:

$$H_2 + \tfrac{1}{2}O_2 \rightarrow H_2O$$

Therefore, the fuel cell produces water that accumulates at the cathode. This product water must be continually removed to facilitate further reaction. See **membrane-electrode assembly**.

Polyurethane elastomers

These materials, applications of which are numerous, play two major roles in fuel technology. First, they are used to insulate and protect chemically oil- or gas-bearing pipes. Secondly, they are used to coat coal-handling plant to minimise erosion of metal surfaces due to the movement of abrasive lumps of coal over them.

Poor man's fuel

In countries including India many homes are not connected to gas or electricity and kerosene is used for illumination, heating and sometimes cooking. There is organised distribution of kerosene to such homes and kerosene so supplied is often referred to as 'poor man's fuel'. In fact, about half of the kerosene released by the refineries for such distribution is diverted to black markets, where its subsequent uses include illegal blending with motor fuels.

Port Arthur, TX

Scene of several refineries including one close to the **Spindletop field**, established only two years after the famous gush in 1901 at Spindletop. This particular refinery was a major supplier of high-octane fuel for piston-engined aircraft used by the USAF during WW2. It is now operated by **Motiva Enterprises**.

Port Dickson, Malaysia

Location of a major oil refinery operated by Shell with a capacity of 125 000 **bbl** per day. Its output was enhanced, not only in quantity but also in selectivity of products, by investment a few years ago in a new **catalytic cracking** unit. Among other benefits, this enables propylene to be produced at the refinery.

Port Talbot

A town in Wales, the scene of a blast furnace accident in 2001 when **blast furnace gas**, rich in **carbon monoxide**, was released. Three died and there were seventeen non-fatal injuries.

Port Waratah

On the central coast of New South Wales, Australia, one of the largest coal exporting ports in the world typically handling 7 Mt per month. Most of the coal departing Waratah is bituminous and locally mined at the Hunter Valley. Recently many shipments from Port Waratah to **China** have taken place.

Portugal, paucity of energy reserves

Portugal has no significant reserves of oil or natural gas. Exploration for **deep-sea oil and gas** in Portuguese waters is underway although unpromising results have caused some oil companies to lose interest in applying for extensions to exploration licences. There are two refineries for the processing of imported crude. There is some **anthracite** which was mined in modest quantities and applied to power generation until 1994 when its use became uneconomic. Hydroelectric power provides about 20% of Portugal's electricity needs, the remainder being generated thermally with imported fuels. The **Alqueva project**, completed in 2002, significantly raised the hydroelectric contribution to Portugal's power. **Liquefied petroleum gas** is imported into Portugal in significant quantities, some entering the country at the terminal at **Sines**.

Porvoo refinery diesel project

The Porvoo refinery in Finland is one of the largest in Scandinavia, receiving crude oil from the North Sea and from Russia. In response to an evident shortage of diesel in Europe, the refinery is raising its diesel production by using a slightly wider temperature range at the fractionating stage. The resulting material therefore has a proportion of what would otherwise have been residue, and so requires a degree of hydrogenation to bring it to within the specifications of diesel. The

hydrogen so required will be obtained by **steam reforming** of natural gas and other **hydrocarbon** gases which might become available in the course of the refinery's operations. By the time the diesel plant is running to full capacity it will require about 15 tonnes of hydrogen per hour for sustained operation. The diesel produced, which will be marketed locally, will have a sulphur content below 10 ppm.

Pot furnace

Also known as a combustion pot, a pot furnace is a device for examining the combustion performance of particular coals. It comprises a cylindrical steel pot capable of holding a quantity of the order of a kg of the coal, with an open top and a refractory base. Air from a compressor is supplied at the base and there are a number of thermocouples at various depths in the pot. The coal charge is ignited by placing **charcoal** soaked in kerosene at the top, and applying a flame.

The behaviour most commonly observed with coals other than those particularly low in volatiles, is that the volatiles burn initially. The progress of their combustion, from the top of the pot to the base, can be followed by observing a ring of red heat descending the outside pot wall. Combustion of the residual solid follows and the ring of red heat moves up the pot again. Throughout the burning, at the top of the pot there will be a yellow flame of height about 20 cm, depending on the air supply rate. The behaviour where volatiles and residual solid combustion are distinctly observable is believed to simulate combustion of coal in a travelling grate stoker particularly well. Often the residue remaining after a pot furnace test will be analysed for unburnt carbon.

Accessories and extensions to the pot furnace are common; for example there might be a **chemiluminescence meter** in position to measure oxide of nitrogen release at different stages of the combustion. **Gasification** of coals can also be studied in a pot furnace.

Poultry litter

Poultry litter can be used as a fuel for power generation. Power from such fuel is currently a reality at places including Norfolk, England, where generation is at about 35 MW. The ash residue is high in nitrates and can profitably be blended with soil.

Power from renewables, target for 2020

The European Parliament has expressed a hope that 25% of the power generated in each and every member state will, by 2020, be from **renewable** sources including **wind turbines**, **energy crops** and **photovoltaic**

cells. The same declaration contains target percentages of biofuels in gasoline by that time.

Poza Rica

The location of a large reserve of oil in central Mexico discovered in 1932 and still producing, operated by Pemex. In 1950, by which time Poza Rica was Mexico's most productive oil field, there was a very serious accident there. **Associated gas** very high in hydrogen sulphide ('sour gas') was being diverted to a flare for incineration. The flare malfunctioned and the hydrogen sulphide entered the air, fatally poisoning 22 persons and necessitating the hospitalisation of over 300 others.

Premier Oil

Oil-producing company originally named the Caribbean Oil Company in 1934. Its activities were exclusively in Trinidad until, as the Premier Oil Company, it became involved in North Sea exploration in 1971. It retains an interest in the North Sea and has a significant stake in the **Wytch farm** oil field. In recent years it has been involved in exploration and production in countries including Pakistan, Indonesia and Myanmar. Indonesian activity is particularly strong, the company having acquired Sumatra Oil Ltd. in 1996.

Prestige tanker spill

The oil tanker Prestige broke in half off the Spanish coast in November 2002. The vessel, which was displaying the flag of the Bahamas, had first encountered trouble the previous week during bad weather. About 50 000 **bbl** of oil, roughly 10% of the vessel's payload, was released at the initial breakage and an oil slick 70 miles long resulted with damage to beaches and harm to wildlife. The rear part of the vessel containing many of the oil tanks went beneath the water. It was hoped that the fact that the leakage was in the open sea rather than in a restricted area of water as with **Exxon Valdez** would aid dispersion and mitigate the effects of the spill. Some of the tanks from the rear of the vessel descended intact to the seabed and their contents will be salvageable. The depth—over 3500 m—might make such an operation uneconomic, however.

Primary air

Air supplied and mixed with the fuel on its way to the **burner**. It is supplemented at the flame with secondary air which diffuses into the flame from the atmosphere. For example, in a domestic cook-top, premixed gas and fuel burn to form the inner cone of a flame. The outer cone is formed by excess fuel exiting the premixed flame and burning with atmospheric oxygen. Such a flame is therefore a composite of a premixed flame and a diffusion flame.

Primary cell
A self-initiating, irreversible, electrochemical cell where the chemical energy of the constituents is converted into electrical energy when the current is allowed to flow.

Primary energy
Energy contained in raw fuels and any other form of energy received by a system as input. Primary energies are transformed in energy conversion processes to more convenient forms of energy, such as electrical energy and cleaner fuels, known as **secondary energy**.

Processed natural gas
Natural gas modified by **steam reforming** to give a gas with **calorific value** about the same as that of manufactured **town gas**, that is about 20 MJ m^{-3}. A common application is during a transitional period when natural gas replaces a manufactured gas but **burners** in the supply area have not all been adjusted for natural gas.

Producer gas
Gas made by passing air or air/steam through a hot bed of fuel, often **coal**. The dominant reaction is:

$$C + \tfrac{1}{2}O_2 \rightarrow CO$$

the process being exothermic so that the temperature of the bed is maintained. The plant for making producer gas is called a gas producer. The **calorific value** of such a gas will be in the range 4–5 MJ m^{-3}. This is only an eighth to a tenth of the calorific value of natural gas, yet producer gas can melt steel. Producer gas came into widespread use in the second half of the 19th century and remained a staple fuel in many industries, including the glass industry, for about a century. It was of course recognised that the heat which keeps the fuel bed hot is heat which would have been available to the user if the coal had been burnt directly, therefore gasification in this way incurs an energy penalty. This is offset by the advantage, for some applications, of a gaseous fuel over a solid one.

There has been revival of interest, partly because very cheap feedstocks such as **municipal solid waste** can be gasified in this way. Cellulosic wastes including wood waste, corn cobs and peanut shells have also been used to make producer gas as have coal fines and **coke breeze**. It is possible to power a motor vehicle with producer gas, using a gas producer *in situ* at the rear of the vehicle. Producer gas manufacture is an example of total gasification: all of the organic part of the feedstock is gasified and only ash remains. This contrasts with partial gasification where gas is produced by **devolatilisation** and there are **char** and tars/oils as accompanying organic products.

Currently, producer gas finds application to the powering of **genset** devices in places including India.

Production tax
Either tax per unit of pollution generated, or tax per unit of energy produced.

Prometheus gas project
Underway in the North Sea about 30 miles out from the East Yorkshire Coast, exploration wells are being drilled. It is expected that gas eventually produced will be converted to **liquefied natural gas**.

Prompt NO$_x$
The fixation of atmospheric nitrogen by hydrocarbon compounds in fuel rich areas of the **flame**. For example:

$$CH + N_2 \leftrightarrow HCN + N$$

$$C + N_2 \leftrightarrow CN + N$$

NO is subsequently formed by oxidation:

$$N + O \leftrightarrow NO$$

HCN and CN also react to form NO via a complex mechanism. Prompt NO is formed in all combustion systems but its contribution to total **NOx** emissions is combustion system and fuel dependent.

Propylene, liquefied
The critical temperature of propylene is such that, like propane, it can be liquefied at ordinary temperatures by application of pressure and stored and transported that way. Propylene so transported is usually destined for polymerisation rather than for burning.

In July 1978 a road tanker bearing liquefied propylene overturned near a campsite in Spain. The tanker contents exploded and 140 persons were killed.

Protein foam
A fire fighting agent, applications of which include fires resulting from spillages of flammable liquids. It is prepared for use by blending a liquid protein concentrate with water. The protein constituent is fibre protein similar to the keratin in wool and is sometimes obtained from feathers. A protein foam might also contain a fluorinated agent to improve spreading and penetration.

One simple performance criterion for such foams is the extinguishment application density, that is, the amount of foam required per unit area of the pool for extinguishment, measured in litres per m^2, referred to as the extinguishment application density. The value of this for a particular fire will depend on whether the nozzle releasing the foam is fixed or not. If its direction can be controlled by a trained operator, the foam will be more effective than if the nozzle is fixed, in which case full advantage of the foam will not be obtained.

Proto-petroleum
The dead organic deposition from which petroleum is formed.

Proton exchange membrane fuel cell
See **polymer electrolyte membrane fuel cell**

Proven reserves
Estimated amount of coal, crude oil and natural gas that can be recovered under existing economic and operating conditions.

Proximate analysis
Analysis of a coal or other solid fuel according to four categories of product on heating: volatiles, fixed carbon, moisture and ash. If evaluated by the traditional approach and approved by many bodies including ISO, each component is determined in a separate test using a different sub-sample of the coal. Moisture is determined by heating at 105°C. Volatiles are determined after heating in a crucible at 950°C. Ash is determined by allowing a sample of the coal to burn out totally to constant weight, and fixed carbon is determined by difference. The so-called fixed carbon is likely to contain a not insignificant amount of hydrogen and is not the same as the carbon content determined by **ultimate analysis**. Different authorising bodies might specify slightly different temperatures for the respective parts of the procedure.

Pseudo-anthracite

A coal corresponding to the definition of an **anthracite** in terms of its carbon content but yielding slightly more volatiles than are expected from anthracite, making it unsuitable for some applications. Significant deposits include one in New South Wales, Australia.

Pulau Sebarok Terminal

A facility for oil-bearing vessels in Singapore Port. The storage capacity is 1 035 000 m³ of oil. **Singapore Petroleum Company** are among the several companies utilising the terminal. A major benefit to the terminal is that it has facilities for value-adding. Distillates brought to the terminal can be blended before despatch, arriving at their destination ready for use. Similarly **bunker fuel** can be produced at the terminal by blending fuel oils from a number of sources to give a suitable **viscosity**. Note that guidelines apply to the blending of fuel oils. Contravention of them leads to sludge formation.

Pulverised fuel (pf)

Coal having previously been milled to particle sizes of the order 10–100 μm (usually 75% < 75 μm). Milled fuel and air are admitted together to a **burner** to give a jet flame. Coals of all ranks are suitable for burning by this method, although with **anthracite** high grinding energies are involved. Grinding energies for **bituminous coal** or **brown coal** are lower, and the grinding energy needed to prepare any coal for pf combustion will be determined in advance by measuring its **Hardgrove Index**. A side effect of milling brown coal is that it is significantly dewatered, eliminating the need for any further drying before admittance to a pf burner. Nevertheless, in pf combustion with low-rank coals, flame instability due to fluctuations in the moisture content of the fuel can occur. In this eventuality, the raw brown coal admitted for milling is temporarily replaced by **briquettes**, the moisture content of which is lower and less variable than that of raw coal.

Pumped storage of electricity

Means by which in **hydroelectricity** water previously pumped to the upper reservoir can be made rapidly available to the turbine by opening head gates which function as taps. In this way, power can be increased when there is a sudden rise in the demand for it, for example at times of day when households have their electric cookers in operation. Among the many hydroelectric facilities providing for pumped storage is the Dinorwig facility in Wales, which is said to have the fastest response time of any such facility in the world. Another such facility is in Austria. This country has 11.4 GW of hydroelectric power, 6.4 GW of which is from pumped storage facilities.

Pumpherston, Scotland
The location of production of oil from shale during the period 1851–1962. It was close to the conventional fuel facility at **Grangemouth** where many of those who had previously worked at the shale facility were transferred when it closed down. Its closure signalled the end of the once strong Scottish shale industry.

Pumping jack
Also known as a 'nodding donkey', the familiar rocking part of the well pump, which always seems to appear in pictures of onshore oil production. Sometimes a well will initially release oil under its own pressure but will require a well pump as the internal pressure drops which will usually be a long time before depletion.

PUROX process
Technique pioneered by Union Carbide for the gasification of **municipal solid waste (MSW)**. MSW is admitted to a vertical **pyrolysis** reaction chamber from the top, and oxygen (not air) admitted close to the base. Oxygen reacts with char yielded by the waste and, since the oxidant is pure oxygen, post-reaction gases are very hot. These rise up the chamber and pyrolyse incoming waste, yielding char for further reaction with oxygen and also a product gas of typically 11.1 MJ m^{-3} **calorific value**. The gas is burnt on site to make electricity and there is some to spare which can be sold to the grid. The metal and glass constituents of the waste are melted and, on quenching with water, form a hard slag which is easily manageable.

Pyrite
Iron sulphide, formula FeS_2. It is present as part of the mineral matter of many coals and, under conditions where water is present, can exacerbate the propensity of the coal to spontaneous heating. Pyritic sulphur, like organic sulphur, converts to **sulphur dioxide** on coal combustion.

Pyrolysis
More specifically in reference to biomass or coal, pyrolysis describes the process of producing a solid carbonaceous **char** (including mineral matter), liquid **tars** and gases (CH_4, CO and CO_2) at a high temperature in the absence of air.

Q

Q8

The **OPEC** countries are traditionally suppliers of crude oil which is sold as such and refined elsewhere. The first OPEC country to become involved in downstream activity was Kuwait, and the associated company was given the homophonic name 'Q8'. Consultants from outside Kuwait were involved in establishing Q8, which since its creation in 1985 has established itself as a major retailer of gasoline, having 5000 service stations. It is also a supplier of jet aircraft fuel, having secured a contract to supply Krabi Airport in Thailand with its fuel needs.

Qatar, hydrocarbon reserves of

This **OPEC** country has 600 Mt of crude oil and **natural gas liquids (NGL)** have been produced since 1949. Current production is ≈700 000 **bbl** of oil per day, about a tenth of the quantity of NGL, reflecting the fact that Qatar is stronger in natural gas than in crude oil. Qatar in fact has 10 900 **bcm** of natural gas, 7% of the entire global reserves. Although some of the oil fields do contain **associated natural gas**, most of the gas produced in Qatar is from non-associated fields including the **Qatar North Field**. Much natural gas from Qatar is exported as **liquefied natural gas** to destinations including Japan and Mediterranean countries.

Qatar North Field

The Qatar North Field is the world's largest **non-associated gas** field off the northeast coast of the country, containing in excess of 20 trillion m³ of gas. There is also condensate, which, once the gas is brought ashore, is stripped off at a facility operated by the UAE-based Dolphin Energy. At present construction is underway on a sub-sea pipeline to export gas from the Qatar North Field to Abu Dhabi.

Qeshm Island

Location of a fairly new (operations began in 2002) Iranian refinery within a Free Trade Zone. The refining capacity is 120 000 **bbl** per day and crude is supplied from the Gurzin and Salakh fields in Iran, both close to Qeshm Island. Crude from each of these is 'sweet'. Being in a Free Trade Zone, the undertaking has attracted many overseas investors from nations including Canada, and at the time of writing there is a strong possibility that China will also invest in Qeshm Island. From early 2006, Qeshm Island has been receiving **associated gas** by pipeline from the **Sirri oil field** in Iran, which previously had been simply flared.

Quality of heat

A term which describes how easily the heat in something can be transferred to something else. In other words, high-quality heat is at a temperature sufficiently high for there to be heat transfer towards something cooler. A joule of **enthalpy** in a medium at 25°C is quantitatively the same as a joule of enthalpy in a medium at 200°C. However, in the 200°C medium it is qualitatively superior in that it is in engineering practice more easily transferred. Sometimes heat is bought as such, there being no transfer by sale of anything tangible. For example, a city incinerator might sell its heat to an organisation which will divert it to such purposes as hot water production for homes (Sheffield, UK, is one site of such activity). Quite clearly one cannot price such heat simply on a per joule basis; the basis of pricing has to be quality as well as quantity.

Quarl

The inside surfaces of the burners consist of tiles which are made from silicon carbide (SiC). This is a refractory material, which, at high temperatures, has dimensional stability, high thermal conductivity, abrasion, corrosion and thermal shock resistance. When exposed to air at low temperatures, a thin layer of silica (SiO_2) forms on the surface of the SiC tile, providing protection from further oxidation under **burner** operating conditions. These tiles are placed at the mouth of the burner, known as the quarl, in the coal-fired power station. The temperature profile across the tiles is high with the back of the tile at 450°C, the side facing the furnace at 1500°C and the side facing the flame in the burner mouth at 1200°C. In US terminology, the word tile is used for quarl.

Queensland State Gas

A utility receiving and selling on **coal bed methane**. It comes from the mine at **Moura** and, after moisture removal, is combined with gas from other, conventional sources for supply to the homes and businesses of Queensland.

R

Rabigh

Located on the Red Sea coast of **Saudi Arabia**, Rabigh is the location of a projected petrochemicals complex which when complete will be one of the largest in the world. It will receive over 2 Mt of **olefins** per day for processing. The Sumitomo Chemical Company of Japan have a major interest.

Rail cars, for liquefied natural gas (LNG)

These use membrane insulation as do the most up-to-date ocean tankers for carrying this substance. A typical capacity of an **LNG** rail car is 120 m^3, equivalent to just under 50 tonnes of LNG. Such a railcar supplied new will attract a price of about half a million Euros.

Rain forests, decline in carbon sequestration caused by reduction in

In the rain forests of the world an estimated 1050 million trees per year are felled for timber. This can be shown to equate to a loss of ≈25 Mt yr^{-1} of carbon sequestration capacity.

Raniganj coal deposit

The first coal mine in India, in Western Bengal in 1774. The deposit was appropriated by the Bengal Coal Company in 1843 and a geological survey a few years later revealed that the deposit was much more extensive than had originally been realised. The coal is bituminous in rank and across the extensive acreage of the deposit varies in suitability for coking. It is low in sulphur but moderately high in ash.

Coal production at Raniganj continues. Since 1975 the deposit has been owned by Eastern Coalfields Ltd., a subsidiary of Coal India Ltd. Raniganj coal is exported to Bangladesh, Nepal and also to Bhutan in the Himalayas. This is the only coal exported from India. The Raniganj deposit is also seen as a promising source of **coal bed methane**.

Rankine cycle

A thermodynamic cycle, usually analysed in temperature-entropy terms, in which the working substance is steam. The Rankine cycle is the basis of a great deal of thermal generation of electricity. In the simplest form of the Rankine cycle, saturated steam is allowed to cool and, to the extent that the process is reversible, the entire energy effect is manifest as work. Performance enhancement by superheating the steam is possible and this is very common in power generation. Other ways of improving performance include adjustment of condensation conditions at the end of the work-producing step of the cycle.

In a Rankine cycle the role of the fuel is simply to raise the steam which itself undergoes the cycle. Any available fuel—solid, liquid or gaseous—can be used.

Rankine temperature scale (°R)

The absolute form of the Fahrenheit scale. Absolute zero i.e. the temperature at which molecular energy is a minimum, corresponds to $-273°C$ on the Centigrade scale. On the Fahrenheit scale, absolute zero or $0°R$ is equivalent to $-460°F$, hence:

$$°R = °F + 460$$

Rapeseed

A source of **biodiesel**. In the European Union, 10 Mt are grown annually, not all of which is diverted to fuel use as there is some utilisation as cooking oil. In warmer climates, a comparable product to rapeseed oil is produced from the **Jatropha tree**.

Ras Laffan, Qatar

Scene of a proposed ethylene plant expected to begin production in 2012. Participants are ExxonMobil and Qatar Petroleum and the plant when complete will have polymerisation as well as cracking facilities.

Ras Tanura, Saudi Arabia

Scene of the world's largest oil terminal, the first shipment having been in 1939. It has eleven berths for oil tankers, divided between the north and south piers of the terminal, and currently operates at a capacity of 6 million **bbl** per day. There are facilities for transferring from shore to ship not only crude oil (there is a major refinery in Ras Tanura) but also refined material and **liquefied petroleum gas**. About eighteen miles along the coast and seven miles out to sea is the Juaymah crude oil terminal which has a capacity of about 3 million bbl per day. In fact Ras Tanura and Juaymah sometimes feature in news accounts (and on the web sites of engineering companies active in the area) as if they were a united facility. All crude oil despatched from Ras Tanura or from Juaymah has passed through the processing facility at **Abqaiq**.

Raw fuel

A fuel source such as coal, natural gas or wood that is used for energy in the same form in which it is found in nature, with little or no processing.

Rayong Refinery

Grass roots refinery in Thailand close to the southern border of the country, with a production of 145 000 **Barrels per stream day (BPSD)**. The facility is self-sufficient in electricity having a power plant producing at almost 100 MW.

Recuperative burning

Burning such that sensible heat from the combustion products is transferred by heat exchange to the incoming fuel and air, raising their temperature. By this means, temperatures in excess of the usual adiabatic flame temperature (based on entry of reactants at room temperature) can be exceeded. In recuperative burning, heat exchange is across a solid boundary distinguishing it from **regenerative burning**.

Recycling

The process of recovering or extracting useful or valuable materials from waste. For example: following their useful life, lead acid batteries can be recycled. The polypropylene battery case can be turned into pellets; battery acid into high purity gypsum; lead into pure or alloy ingots, blocks, extruded lead wire or shot. Modern recycling plants can achieve up to 99.7% lead recycling from **lead-acid batteries**.

Red Hawk
The name of a fairly recently developed field in the Gulf of Mexico currently productive of gas only. It is expected to be yielding 3 million m³ of natural gas per day once fully operational. It is a deep field ($\approx$5000 ft) and the production platform in use is a floating one.

Refinery
An industrial facility for refining crude oil. Crude oil is composed of different hydrocarbons which can be separated by distillation since they have different boiling points. Since the lighter liquid compounds are in great demand for use in internal combustion engines, a refinery will convert heavy liquid and lighter gaseous elements into these higher value products using complex and energy intensive processes. Following separation and removal of any impurities, the distillation petroleum products can be either sold without any further processing, or passed on to other processes such as **catalytic cracking**, **hydrocracking**, **hydrotreating** or **visbreaking** to produce a range of value-added products. The American Petroleum Institute have published figures for the products from a typical barrel of oil: 19.5 gallons gasoline, 9.2 gallons distillate fuel oil, 2.3 gallons residual fuel oil, 1.9 gallons liquefied gases, 1.9 gallons still gas, 1.8 gallons petroleum coke, 1.3 gallons asphalt, 1.2 gallons feedstocks, 0.5 gallons lubricants, 0.2 gallons kerosene and 0.3 gallons of other products.

Reflectograms
Show the reflectance distribution in the form of a frequency histogram. They may be constructed on whole coals, blends, or an individual **maceral** group. The main use of a reflectogram is to analyse and monitor the quality of coal blends as they provide a means to distinguish between different ranks within the blend. Normal chemical analysis will only give a mean value of rank for the whole mixture. Reflectograms can be difficult to interpret when the blend components are of a similar rank and the method is not unambiguous. Reflectograms are best confined to coals of medium rank. See **vitrinite reflectance**.

Reforming
Reforming is a chemical process that reacts hydrogen-containing fuels in the presence of steam, oxygen, or both into a hydrogen-rich gas stream. When applied to solid fuels the reforming process is called **gasification**. The resulting hydrogen-rich gas mixture is called reformate.

The equipment used to produce reformate is known as a reformer or fuel processor. The specific composition of the reformate depends on the source fuel and the process used, but it always contains other compounds such as nitrogen, **carbon dioxide**, **carbon monoxide** and some of the unreacted source fuel. When hydrogen is removed from the reformate, the remaining gas mixture is called raffinate.

Reforming a fossil fuel consists of the following steps:
• feedstock purification (including sulphur removal)
• steam reforming or oxidation of feedstock to form hydrogen and carbon oxides
• primary purification, the conversion of carbon monoxide to carbon dioxide
• secondary purification, further reduction of carbon monoxide.

The advantages of reforming fossil fuels are that they:

• use existing fuel infrastructures
• reduce the need to transport and store hydrogen
• do not need the input of large amounts of energy as in electrolysis
• are less expensive than other hydrogen production methods.

The disadvantages of reformers are that they:

• can have relatively long warm-up times
• are difficult to apply to vehicle engines because of irregular demands for power
• are complex, large and expensive
• introduce additional losses into the energy conversion process, especially those that have small thermal mass
• use non-renewable fossil fuels
• generate pollution.

The pollution generated by reformers take three forms: carbon dioxide emissions, incomplete reactions leaving carbon monoxide and some of the source fuel in the reformate, and production of pollutants through combustion, such as **NOx**.

Reforming fossil fuels only makes sense if the hydrogen is needed directly, as in a fuel cell engine. For internal combustion engines, it is always more efficient to use the fossil fuel directly without passing it through a reformer first.

Reformers are of three basic types: steam reformers, partial oxidation reactors and thermal decomposition reactors. A fourth type results from the combination of partial oxidation and steam reforming in a single reactor, the autothermal reformer.

Refractory materials

Materials with a high degree of resistance to heat, used in the construction of a **firebox**. They are required to have high softening temperatures and also chemical inertness so that they will not be affected by gaseous combustion products such as **sulphur dioxide**. They are also required to bear a considerable weight of fuel and this can cause failure at temperatures below the softening temperature in the absence of such a load. Common refractories include fireclay brick which, depending on the grade, can be used at temperatures up to about 1750°C. The two most common forms of firebrick failure are spalling and slagging. Spalling involves disintegration of the firebox inside surface, comprising the refractory exposing new surface. In planning its avoidance the thermal conductivity of the refractory is a factor. Slagging is reaction of the refractory with the ash-forming constituents of a solid fuel. See **quarl**.

Regasification plant

Term applied to a device used to vaporise **liquefied natural gas (LNG)** other than by heat exchange with water. Sometimes when natural gas used for power generation is delivered as LNG and there is a **combined heat and power** cycle, some of the heat is used to vaporise (regasify) the LNG. The Dragon Natural Gas facility in Wales, occupying the site of a former refinery, is an example of such an enterprise.

Regenerative burning

Burning such that hot post-combustion gas is mixed with cold influx gas in order to raise its temperature and that of the subsequent combustion. Regenerative burning finds application in steel processing.

Regenerative fuel cell

A fuel cell in which the reaction products are recycled to form reactants. For example, water (H_2O) produced by an electrochemical process is separated into hydrogen (H_2) and oxygen (O_2), which are then fed back into the fuel cell.

Renewable energy

Renewable energy is energy that is regenerative or, for all practical purposes, virtually inexhaustible, for example: solar energy, wind energy, hydropower, plant-derived **biomass**, geothermal energy and ocean energy. The increased use of renewable energy could reduce the burning of **fossil fuels**, eliminating associated air-pollution and carbon dioxide emissions.

Renewable energy makes important contributions to world energy supplies. Hydroelectric power is a major source of electrical energy in many countries, including Brazil, Canada, China, Egypt, Norway, and Russia. In developing countries many people do not have access to, or cannot afford, electricity or petroleum fuels and use biomass for their primary energy needs. For example, most rural people in Africa use wood, scrub, grass, and animal dung for cooking fuel. Small-scale renewable energy technologies are often the only practical means of supplying electricity in rural areas of these countries.

Renewable energy, growth in UK
Electricity generated from renewable sources was 3.6% of the total in 2004 compared to 2.7% in 2003, representing 1.7 GW and 1.22 GW respectively. The **wind turbine** contribution reached 1 GW for the first time in August 2005.

Reservoir
The predominant reservoir for hydrocarbon accumulations in global environments is in sandstones. The origin of the reservoirs can vary from deltas to deepsea turbidites, for example, the North Sea and Nigeria.

The predominant reservoir for Middle Eastern reserves in the Arabian Gulf and the Mesopotamian Basin is limestone. This is also the significant reservoir for reserves in Mexico, Western Canada and various basins in the United States.

Fractured basement rocks, fractured cherts and fractured volcanic rocks also act as reservoir rocks.

Resource
Anything that has value for use. Coal, oil, natural gas and water are all natural resources.

Retort coal gas
An example of a **town gas** manufactured by heating coal to give, in addition to the gas, tars/oil and **coke** as by-products. Retort coal gas has a **calorific value** of 20 MJ m^{-3} and is similar in composition to coke oven gas. In a centre such as Pittsburgh where a great deal of coking takes place to service the steel industry there is coke oven gas to spare, which can be supplied to users interchangeably with retort coal gas.

Retrofit

Modification of an existing plant. For example: addition of a flue gas desulphurisation plant on to an existing coal-fired power station.

Reverberatory furnace

A reverberatory furnace is a metallurgical or process furnace which isolates the material being processed from the fuel, but not from the combustion gases. Process chemistry determines the optimum relationship between the fuel and the material, among other variables. Some applications require contact between the material and the hot gas. Contact with the products of combustion, which may add undesirable elements to the subject material, is used to advantage in some processes. Control of the fuel/air balance can alter the exhaust gas chemistry toward either an oxidising or a reducing mixture, and thus alter the chemistry of the material being processed. The term reverberatory is used to mean rebounding or reflecting. Often abbreviated to 'reverb furnace'.

Rhine region

The location of the largest formation of **brown coal** (also referred to as **braunkohle** or **lignite**) in Europe.

Rhum field

Non-associated gas field in the North Sea, known to contain 20 million m^3 of gas and developed by BP and the National Iranian Oil Company; production commenced in 2005. There are at present three wells at Rhum and gas from them is taken by pipeline to the production facilities at the **Bruce field**. Rhum gas has the disadvantage of a fairly high carbon dioxide content, which has made previous development of the field uneconomic. Existing production capacity at the Bruce field is sufficient to take gas from additional wells at Rhum if and when these are drilled.

Rice husks

Waste material comprising particles of a fairly regular shape with potential use as a solid fuel. Cellulosic in composition, they can be used as a fuel for direct burning (for example in parts of Malaysia and in Vietnam) or as a gasification feedstock (for example in the Riverina region of Australia).

Richards Bay

Location in Kwazulu-Natal, South Africa, and the scene of one of the largest coal exporting facilities in the world, a few miles along the coast from Durban. South Africa exports about 75 Mt yr^{-1} of coal, most of which passes through Richards Bay. The port has been handling coal since 1976 and there are plans underway to extend the capacity by 10 Mt yr^{-1}.

Ridley Island

Scene of a deepwater, ice-free harbour off British Columbia where, since the 1980s, there has been a terminal for coal exports from BC. Initially the property of the Canadian Government, it was recently sold to the UK-based Fortune Minerals who have coal winning interests in BC. The capacity of the terminal is 16 Mt yr^{-1}, but in 2004 only 1.3 Mt were handled, leading to heavy financial losses. This is expected to improve partly because of the increased demand for imported coal in China.

Road tankers, for liquefied natural gas (LNG)

These use membrane insulation for the tank itself and during a journey lasting three days, losses of the payload will be negligible. Such a tank will be capable of carrying about 50 m^3 of **LNG**. The tank has features to protect it against breakage in a collision. Sometimes the tank also carries facilities for vaporisation of the LNG on delivery.

Rock Rive Wind Farm

Situated in Wyoming, this facility has been producing power since 2001. It has fifty 1 MW **wind turbine**s, manufactured by Mitsubishi. Wyoming has in fact been identified as one of the most promising states in the US for wind farms with the potential to provide the region with 50 000 MW of power. Another wind farm development in a northernmost state of the US is **Pierce County, ND**.

Rocket

Device capable of moving at supersonic speeds by passage of the product gases from the fuel through a nozzle. The high speeds of the gases are brought about by conversion of **enthalpy** to kinetic energy. This is distinct from a jet engine which is a gas turbine involving a thermodynamic cycle.

The thrust of which a rocket is capable bears an inverse relationship to the molar weight of the product gas, hence rocket fuels which produce low molecular weight chemical species such as water are preferred. **Hydrogen peroxide** is an example. Not only a **monopropellant** but a conventional fuel, it can be used to power a rocket. For example jet fuel, comprising **kerosene**, is sometimes used in rocketry in which case liquid oxygen will be carried for use as oxidant (or perhaps an alternative oxidant such as fluorine). Carbon dioxide, the major **combustion** product from kerosene, has a relatively high molecular weight (almost two and a half times that of water) and this has an adverse effect on the thrust.

Rockingham

A site in Western Australia (WA) where **landfill gas** is used to generate power. Two spark-ignition **gensets** are employed; one capable of generation at 1 MW and the other at 0.6 MW, each with a frequency of 50 Hz. Power so produced is exported to the grid. Also in WA is the Gosnells landfill site where two **gensets** each of 1.05 MW are in use.

Rocky Mountains Region, gas reserves of

The natural gas reserves are 5 Tm^3, much of it untapped as yet. In addition there is an estimated 1 Tm^3 of **coal bed methane** awaiting developments whereby it can be added to the US natural gas system. The US imports about 20% of its natural gas from Canada. The development of coal bed methane reserves in states bordering with Canada might significantly reduce this import requirement. The Powder River coal deposit in Wyoming is seen as one of the most promising sources of coal bed methane in the Rockies, and tens of thousands of drilling sites are expected there within the next decade.

Rolls Royce RB211, use of in offshore oil and gas production

This gas turbine, which for several decades has been used for jet aircraft, has also found application in offshore activity. An example is a support platform topside located in the **Caspian Sea** where it will service three offshore platforms. The support platform contains ten RB211 turbines. Two will be dedicated to electrical power production and others will be used as necessary for water and gas injection into sub-sea wells.

Romania, oil reserves in

As early as 1860 (the year following the drilling of Drake's oil well) Romania produced its first oil and there was growth to the million **bbl** per day scale by the mid-60s. The Romanian oil fields were appropriated by the German occupation during WW2 and supplemented liquid fuels made from **braunkohle** largely by **I G Farben**.

Roncador field

Oil and gas field in Brazilian waters, and the scene of one of the worst recent offshore accidents when in 2001 the P-36 semi-submersible sank and 11 lives were lost. The field was discovered in 1996 as a result of a **wildcat** drilling operation and is at the present time one of the deepest offshore oil reserves to have been accessed, drilling having been at 1800 m (5900 ft).

Rosneft

Russian oil company having operated there since the reign of Tsar Alexander the Second (1855–81). Rosneft drilled the first oil well in Russia in 1866, seven years after the **Drake** well in Pennsylvania. Its current activities include exploration in the **Sea of Azov**. Rosneft are currently collaborating with the US division of the French company Schlumberger in the development of the **Vankor field** and several other western companies, including Canadian **SNC-Lavalin,** will also be involved.

Rotterdam, wind farm at

Rotterdam is close to one of the densest areas of hydrocarbon processing in the world, namely the Rijnmond region. Consistent with the principle that energy from renewables should, at this stage in history, accompany and complement that from hydrocarbon sources, there is in fact a 22.5 MW wind farm at Rotterdam partly owned by BP. Shell are also active in wind power in the same part of the world at **Egmond aan Zee**.

Other oil companies are, at several of their refineries, investigating wind power for in-house use: Chevron operate a wind farm at **Nerefco**. There is the argument that the refineries are at locations already occupied by industrial plant where the turbines would add nothing unsuitable to the landscape.

Rover Jet 1

The world's first gas turbine powered car, making its debut in 1950. Trials of it did not augur well for subsequent release of such a car onto the market partly because fuel consumption was high and accelerator depression unresponsive. The body was basically that of the conventional Rover 50s and early 60s models, sometimes pejoratively referred to as the 'Rover Auntie'. Rover continued research into gas turbine cars for as long as it was an independent company. US manufacturers also experimented and a 1954 Chrysler Plymouth was converted to gas turbine power. Fiat performed similar trials on one of their models at around the same time.

Gas turbine cars for general use were a failure but a revival of interest is expected now that the **microturbine** has become available.

Ruhr coal field

The site of **hard coal** production, accounting for over three-quarters of that produced in Germany. Both **anthracite** and **bituminous coal** are found there. The bituminous coal from the Ruhr field is suitable for **coke** production.

Russian Arctic oil spill

Occurring in 1994, this was due to pipeline failure near Usinsk in the north of Russia. The pipeline was leaking steadily over a period of some eight months but no action was taken, there being reliance on a dike which prevented leaked the oil from spreading. The dike failed, however, and about 700 000 **bbl** of crude oil were released into the environment. The vegetation of the local terrain was threatened as consequently were the reindeer that feed on it. Reindeer are reared for their meat in that part of the world, so the industry was impacted by the oil spill.

Russia loses about 20% of its annual crude oil production through pipeline leakage. The production for 2005 was approximately 3×10^9 bbl, so if a value of $US 50 per bbl is adopted, the wastage in cash terms is $US 30 billion. The loss in terms of environmental effects, as far as it can be expressed in financial terms, will be much higher still. A semi-quantitative comparison with the costs associated with the **Exxon Valdez** (EV) spill is presented below.

Exxon Valdez: Quantity of oil = 1.1×10^7 bbl

Payout by Exxon, comprising compensation to fishermen, State and Federal settlements and cleaning up costs = $US 3.5 billion

Annual Russian loss: Quantity of oil = $0.2 \times 3 \times 10^9 = 6 \times 10^8$ bbl

If the cost is scaled using the EV figure:

$US 3.5 billion $\times 6 \times 10^8 / (1.1 \times 10^7)$ = $US 190 billion

---

We can cautiously conclude that if environmental protection was as stringent in Russia as in North America, leakage of oil would require payout by the Russian oil companies of the order of $US 200 billion per annum.

Russo-German Pipeline

Proposed pipeline, which for much of its 600-mile length will be submerged in the Baltic Sea, thus by-passing Ukraine and Poland thereby improving security of supply. Expected to come on-stream in 2010, it will supply Russian natural gas to Western Europe, some of which will be for petrochemical use in Germany by BASF.

Ruwais, Abu Dhabi

Scene of a major olefins plant jointly operated by Borouge, a concern in which both the Danish Borealis and Abu Dhabi National Oil Company each have a holding. The plant produces 600 000 tonnes each of **ethylene** and propylene per year and there are plans under way to increase this output by a factor of three by the end of the decade with financial advice from HSBC. It is expected that the eventual polymer resulting from the increased olefin production will be used in the manufacture of insulation for electrical cables manufactured in the Far East.

Rwanda, use of wastes as fuel in

This African country has a scheme underway making **briquettes** for use as a general-purpose solid fuel from **municipal solid waste (MSW)**. The target quantity is 17 885 tonnes of such briquettes per year. At present in Rwanda wood fuels are widely used, since there is only a gas supply to less than 1% of homes. The result is deforestation to a very high degree. The waste-derived briquettes will save a significant number of trees and slow down the depletion of wood fuel reserves in Rwanda.

The pilot scheme has taken place in the Rwandan city of Kigali. In such a place, aluminium cans and glass bottles will not be numerous with the result that a greater proportion of MSW will be combustible than would be the case in a developed country. In fact, it was found in the pilot scheme that 75% of the waste collected was suitable for incorporation into briquettes. During the pilot scheme, the cost at which a manufactured briquette could be sold was found to be 53% of that of wood fuel producing the same amount of heat. The project is now in growth and early sales have been largely to institutional kitchens, for example in schools, in the hope that this will open the door to a wider market.

S

Saar
A coal field in Germany second in productivity of **hard coal** only to the **Ruhr coal field**. The coal is bituminous in rank.

Sable project
Exploratory activity for natural gas, located off the coast of Nova Scotia 140 miles out to sea from Halifax. Provable reserves of gas are about 100 **bcm**. When the fields are producing gas at the projected rate, there will be as a bonus ≈11 000 **bbl** each of **natural gas liquids** and condensate per day.

Interest in hydrocarbon reserves in waters to the east of Canada is recent: the **Hibernia** field, off the Newfoundland coast, is also newly exploited.

Safflower
Crop grown for its petals in areas such as India, used as a colouring agent for foodstuffs and as an ingredient of herbal teas. At an installation currently operating in India, **biomass** residue from the cultivation of safflower is taken to a gasifier where it yields a gas capable of sustaining a flame with 800 kW heat release rate and a char by-product also with some commercial value.

Sago Mine, WV
Bituminous coalmine where, on the second day of 2006, there was an explosion which trapped thirteen miners for nearly two days. Television viewers around the world were told that twelve of the thirteen miners had survived although in fact only one survived and twelve perished.

Sahara desert, coal underneath
There are believed to be a billion tonnes of **bituminous coal** underneath the Sahara desert. At a point in the desert less than six miles from a railway and less than 100 miles from the Algerian port of Oran, production is scheduled to begin. The Moroccan and Spanish markets will be targeted.

Saih Nihayda, Oman
The site of a recently developed non-associated gas field from which 3.3 Mt yr^{-1} of **liquefied natural gas** are produced. Oman is expanding as a producer of gas. For example, the onshore field at Kauther in the north of the country, first discovered in 2001, is also being developed and production is expected to begin in 2008. Gas from the Kauther field is rich in condensate.

Sakata Port, Japan
The location of power generation from **marine wave energy** by **oscillating water column**, currently yielding 60 kW. It is expected this will be increased to 200 kW.

Sakhalin Energy
Project off the east coast of the former Soviet Union, north of Japan, for the winning of natural gas and its conversion to **liquefied natural gas (LNG)**. Partners in the undertaking include Shell, Mitsubishi and the government of Russia. Potential customers for the LNG include Tokyo Electric and Korea Gas. There have been delays, and delivery of the LNG is expected to commence in mid-2008, having initially been scheduled for November 2007.

Salah gas reservoir
A major reserve in Algerian desert terrain and, since 2004, the scene of **carbon sequestration**. One million tonnes of **carbon dioxide** are admitted to wells at this field each year by BP.

Salalah, Port of

Oman port of initial capacity 2.24 million **twenty-foot equivalent units (TEU)** per annum, scheduled to be expanded to 4 million TEU per annum or more. The port, which was opened in 1997, is not solely for hydrocarbons; its other arrivals include foodstuffs. However, at its general cargo terminal there is a berth operated by three oil companies, one of which is BP where refined petroleum products including diesel and gasoline are received and pumped to storage tanks outside the terminal.

Salt domes

Geological feature indicative of the presence of oil. In 1924, a geologist in Texas located a salt dome containing oil in what was the first application of sound waves to such exploration activity. The **Strategic Petroleum Reserve** is held in salt domes.

Salto Grande

The scene of **hydroelectricity** at almost 1000 MW level in Uruguay. There are three smaller such installations in Uruguay with a combined output of about 600 MW, the remainder (about a fifth) of the electricity supply of that country coming from thermal stations. Uruguay relies on hydroelectricity because it has no oil, gas or coal of its own: all such fuels for use in the country are imported. There is a refinery at La Teja which processes about 50 000 **bbl** per day of imported crude oil.

San Diego, proposed use of wind power in

The San Diego Gas and Electric Company intend that by 2010 20% of the electricity it supplies will be from renewable sources. Subject to the approval of the California Public Utilities Commission, the company will build a wind farm with a capacity of 205 MW, the first output from which will be in 2007 or 2008. The Company is also involved in power from **photovoltaic cells** and in **geothermal electricity generation**, each of which will be required to meet the 2010 target.

Sandhills-2

The most recent site in the UK to have been 'spudded' i.e. at which drilling has been initiated, located on the Isle of Wight. The background to the project is interesting. When the adjacent Sandhills-1 well was being explored 25 years ago, there was an unexpected encounter of oil part of the way down to the target depth. No follow-up to this discovery was made at the time, and Sandhills-1 itself was deemed uneconomic to develop. Recent tests have revealed promise that drilling at Sandhills-2, which will be directed at the oil which the drilling at Sandhills-1 intersected, could be developed into a production well. The field containing both Sandhills contains 10–15 million **bbl** of oil. Drilling at Sandhills-2 will be at a depth of 1270 m.

Sanha liquefied petroleum gas (LPG) floating production, storage and offloading (FPSO)

The composition of **liquefied petroleum gas (LPG)**, which naturally consists of propane and butane, can be adjusted. Alternatively, pure propane or **butane** can be derived. At an installation in the Sanha field off the coast of Angola, LPG from two production platforms is passed along to the Sanha FPSO for composition adjustment while still offshore and the products taken ashore by tanker. This facility began operations in early 2005, the FPSO having been constructed in Japan. Development of the Sanha field continues and it includes recovery of the associated natural gas. Flaring of the gas is seen as being wasteful.

Sarawak, oil production in

There were high levels of oil production in Sarawak in the inter-war years, and onshore production was extended into shallow waters during the 1930s. There was at that time a large (mainly British) expatriate population. Production continued but by the 70s Sarawak oil fields were seen as being in decline. During the opening years of the 21st Century, there has been a revival of interest and activity in hydrocarbon production in this part of the world. There are currently a number of appraisal wells in water ranging in depth from 138 ft to almost 5000 ft.

Sasol

Although a multinational company, the name is usually associated with the plant near Johannesburg. Coal from the deposit at **Twistdraai** is converted into liquid fuel via **synthesis gas**. In association with Chevron, Sasol are also involved in the manufacture of **gas-to-liquid fuels**.

A litre of motor fuel from the Sasol plant sells in South Africa for a sum equivalent to about $US 1.

Saturated hydrocarbons

Organic compounds containing carbon and hydrogen where the carbons are attached to each other via single bonds. Straight and branched chains or cylic with no double, triple or free bonds; see **paraffins**.

Saudi Arabia, importance of to world oil supply

Saudi Arabia has massive oil reserves, but there is a less obvious reason for its strategic importance in oil supply. Supply is finely balanced in a world where most industrialised nations are either net exporters or importers of oil and Japan and South Korea have little or no domestic oil. Saudi Arabia is the only country with both the reserves and the flexibility of production rates to respond to a sudden major shortage. Political ramifications are enormous and highly complex.

Savannah

The world's first nuclear powered passenger/cargo ship, entering service in 1962. (The first nuclear powered surface vessel was, however, the Russian icebreaker *Lenin*, in service from 1959 to 1989). It had a 74 MW reactor, which, via a steam cycle, delivered 15 MW (20 000 hp) of mechanical power for propulsion. The vessel was in service for about ten years without ever paying its way, a Federal subsidy always being required. It is now in permanent storage in Virginia.

That the vessel was built was the proposal of President Dwight D Eisenhower, one of whose preoccupations was the constructive use of nuclear power at a time when Hiroshima and Nagasaki were only in the recent past. The **Arco, Idaho** initiative was also part of Eisenhower's plans to use nuclear energy for peaceful purposes.

Savannah, SC

Nuclear material from weapons in the US has been sent to France for conversion to nuclear fuels for civilian use, such conversion is also taking place in the US at Savannah. **Mixed oxide (nuclear) fuel (MOX)** so produced will be used at the **Catawba Power Station** in SC and also at the McGuire Power station in NC. Although the weapons-grade fuel will be brought to Savannah from many parts of the US, all the MOX produced from it will go to Catawba or to McGuire.

Sawdust, co-firing of with coal

Taking place at Tilbury in the UK among other places, this saves coal but more importantly is favourable in **greenhouse** terms as the sawdust, unlike the coal it replaces, has the quality of **carbon neutrality**. Tilbury power station generates electricity at up to 1020 MW and at present about 5% of the coal otherwise required is replaced by sawdust. Sawdust is of course classed as wood waste, not fuel wood.

Scale

Term given to deposited solids in steam generating plant due to inorganics previously dissolved in the feed water. Its primary constituents are calcium and magnesium salts. It has the disadvantage of raising the thermal resistance of parts of the plant on which it has deposited, adversely affecting efficiency. Prior softening of the feed water is the primary means of controlling scale.

Scarborough field

Non-associated gas field off northwest Australia. There are plans during the remainder of the first decade of the 21st century to bring into being a trade route for **liquefied natural gas** from Australia to the west coast of the US. At the time of writing, there is no export of Australian natural gas to the US. Gas from the Scarborough field will be among that destined for the US where, if the proposals are fulfilled, it will be received at an offshore terminal 20 km from the southern Californian coast.

Schiehallion field

Oil field in the North Sea where a platform operated by BP produces 120 000 **bbl** of oil per day. A fire occurred there in mid-2005. There was some disruption to production, but no deaths or injuries to persons.

Scorpio

One of the earliest mobile offshore production facilities, first commissioned for the Zapata Offshore Company (the president of which was George H W Bush, later to become US President), Houston in 1956 for use in the Gulf of Mexico. A second such facility followed in early 1957 and both survived **Hurricane Audrey** later that year. There were ten in use by 1959, not all of them in the Gulf.

Scotford Upgrader

Facility in Alberta, Canada, for upgrading **bitumen** from tar sands to **syncrude**. Shell has a refinery close by and some of the product of the Upgrader is passed along there and also to the Shell refinery at Sarnia. Some however is sold to other oil companies including Chevron.

Sea coal

(1) An old name for **bituminous coal** derived from the fact that before the proliferation of road transport, mines close to the sea were favoured for development so that ships could be used to distribute coal

around the national coastline. It is recorded that such a ship became stranded on the beach in Bournemouth, on the south coast of England. It was re-launched by about six volunteers at a time when the tide was favourable.

(2) Term still in use for coal washed up from deposits below sea level.

Sea of Azov
Connected to the Black Sea and in the region known as the Crimea, the Sea of Azov has as its western boundary the Ukrainian coastline. There is currently much exploration for gas and oil in the Sea of Azov with the Ukraine holding a major stake.

Seabulk Pride
A **double-hulled tanker** which, while taking on refined material at a terminal in Alaska in February 2006, was hit by a floating sheet of ice. The vessel's moorings were broken by the collision and some of the material being transferred from the refinery to vessel leaked onto the deck of the vessel. A quantity of 10 gallons of liquid hydrocarbon found its way into the sea.

Seal rock
An impervious rock above an oil or gas reservoir, preventing further upward migration and allowing the oil and/or gas to accumulate. Without the seal the oil and gas would seep away from the accumulation. The seal is therefore an impermeable barrier that prevents the hydrocarbons from migrating out of the reservoir. Examples of trap rock seals are impervious shales, salt and marls (very fine lime mud). Because of its molecular sizes, gas is more capable of escaping through trap rocks. The most efficient types of trap rock or seal are salt sequences.

Secunda, South Africa
Site of an olefin and surfactants plant operated by **Sasol**. The plant is shortly to be sold to raise capital in order that Sasol can advance its **gas-to-liquid fuels** programme.

Security of supply
The overall goal is to minimise the risks of an unplanned physical interruption in energy supplies. This requires governments to have a range of policies to help bring about uninterrupted energy supply through promoting domestic and international energy market liber-

alisation, diversity of energy sources, international energy dialogue and the provision of timely and accurate information to the market. The physical security of supply needs protection against potential terrorist threats or kidnapping of key workers.

Sedco 703

An example of a **semi-submersible**, which was involved in the discovery of the **Magnus field**. Constructed in the UK and first entering service in 1973, it can descend to up to 2000 ft. It has quarters for 100 persons (and four hospital beds) and storage space for drilling fluid in quantities of thousands of m^3. It is one of a family of Sedco 700 semi-submersibles, the newest of which, the Sedco 714, entered service in 1983, having been constructed in Korea by Hyundai.

Sedimentary rock

One of three main rock types. Sedimentary rocks are formed from existing rocks through processes of erosion, denudation and subsequent deposition, compaction and cementation. The porous and permeable varieties may contain organic matter that ultimately is altered to generate oil and/or gas. The organic matter might have been formed in the rock initially—in which case the rock is a source rock—or it might have migrated there. Exploration for petroleum only ever occurs where there is sedimentary rock. In the prior prospecting, attempts would be made to estimate the depth of the sedimentary rock. Almost two-thirds of oil occurences are in sandstones or related rock types.

SELCHP

See **South East London combined heat and power**

Selective catalytic reduction (SCR)

In an SCR system, vaporised ammonia (NH_3) is injected into the flue gas stream at about 300–400°C and passed over a catalyst. The **NOx** is reduced by the NH_3 to molecular nitrogen.

$$4NO + 4NH_3 + O_2 \rightarrow 4N_2 + 6H_2O$$

$$N_2O + 2NH_3 + O_2 \rightarrow 2N_2 + 3H_2O$$

The catalyst is based on oxides of titanium, vanadium and tungsten. NOx reduction of 80–90% can be achieved. The three different arrangements of SCR in coal-fired plant are:

• High dust: most widely used, installed before particulates removal. Catalysts may, however, degrade by fly ash erosion.

• Low dust: installed after particulates removal. Although this avoids erosion it does require costly hot-side **electrostatic precipitator**.

• Tail end installation: longer catalyst life but flue gas needs to be reheated to operating temperature.

Selective non-catalytic reduction (SNCR)

In an SNCR system, an amine-based reagent (usually ammonia or urea) is injected in aqueous or gaseous form into the flue gas at 900–1100°C, reducing **NOx** to molecular nitrogen.

$$4NO + 4NH_3 + O_2 \rightarrow 4N_2 + 6H_2O \text{ (using ammonia)}$$

$$4NO + 2CO(NH_2)_2 + O_2 \rightarrow 4N_2 + 2CO_2 + 4H_2O \text{ (using urea)}$$

NOx reduction efficiency is 40–50% by SNCR, depending on reagent type and operating conditions.

Seme field

Small oil field off the coast of the African country Benin. It was discovered in 1968 and became productive in 1981. It produces only 50 000 **bbl** annually, all of which is exported.

Semi-submersible

Name given to a type of platform for offshore use which, once it has been towed to a site where drilling is required, can descend to a specified depth. Tanks at the base of the platform (pontoons) are filled with seawater and the platform descends to a depth which can be controlled by the amount of seawater admitted, the remainder of the platform structure remaining above the level of the seawater. The previous generation of such platforms were called submersibles and needed to descend all the way to the seabed for drilling, often there only being about a 10 m clearance above the surface. A state-of-the-art semi-submersible is the Marine 700, manufactured by Marine Drilling Companies. It can be used to drill at depths of up to 5000 ft and is currently being used in the development of the Hoover Diana field and of the Thunder Horse field, both in the Gulf of Mexico. The semi-submersible is not restricted to drilling but can be used for production and also for maintenance work on submerged parts of fixed platforms. An example is the P-36 platform, once in use in the **Roncador field** in Brazilian waters. A semi-submersible in production mode is distinct from a **floating production, storage and offloading (FPSO)** which does not descend into the water below the natural depth due its weight.

Semi-submersibles often see service for tens of years as with the

Sedco 703, and might after one already useful lifespan see rebuild and re-entry into service as is the case with the **Bideford Dolphin**, perhaps one of the best known semi-submersibles in the North Sea.

Senboku, Japan

Location of the largest above-ground facility for **liquefied natural gas (LNG)** storage in the world, owned by Osaka Gas. It will hold 180 000 m^3 of LNG. Japan imports more LNG than any other country, so it is perhaps not surprising that the world's largest storage facility for LNG is also in Japan. Located in Yokohama, it will hold about 10% more LNG than the above-ground facility at Senboku.

There is probably no intrinsic advantage of one extremely large LNG tank over a number of smaller ones having the same capacity in total. However in Japan, land prices are high and this is the motivation for occupying all of the available space with the material being stored in large tanks. This factor is significant in the costing of terminals to receive shipments of LNG. The number of such terminals in Japan is well into three figures.

Senegal

This African country makes major use of **biomass** fuels including **peanut shells**, **bagasse**, **rice husks,** millet stalks and maize stalks. There is also peanut haulm, the stems from peanut crops, at a yield of about 0.8 Mt yr^{-1}. There is a moderate amount of natural gas from offshore fields largely used to make electricity. An oil exploration licence has been granted to **Tullow Oil**.

Senghenydd, Glamorganshire

Scene of a coalmine accident in October 1913 which killed 439 miners. The accident originated with a gas explosion. It was one of three coal mining accidents in the UK in the 20th century to have involved more than 100 deaths. The others were at **Hulton, Lancashire** and **Gresford, Denbighshire**.

Sequestrated carbon dioxide, use of in enhanced oil recovery

This doubly beneficial procedure is currently taking place in the US to the extent of 40 Mt yr^{-1} of carbon dioxide.

Seychelles, energy scene in

This Indian Ocean nation comprising a group of islands imports all of its oil, about 4000 **bbl** per day. Electrical power at 40 MW is generated thermally using imported fuel. There is exploration activity for oil and gas around the islands, nine-year licences having been issued to the Gulf Coast based Petroquest.

Seyler's coal chart

Seyler's classification is one of the outstanding historical classifications of coal based on **ultimate analysis**. Originally presented in 1900, coals were grouped into a series of species according to carbon content, and a series of genera based on hydrogen content. The C and H were corrected for moisture, ash and combustible sulphur. The lowest carbon content was 75% (which excluded lignites) and the hydrogen range was too small to include the H-rich sapropelic coals. He found that all normal humic coals could be placed within a narrow band between 75% < C < 95% and 3.0% < H < 5.7% with a maximum width equivalent to 0.7% H. Any coals above the band were termed 'per-hydrous' (H rich) and those below 'sub-hydrous'. He also found that lines of equal volatile matter ('iso-vols') and calorific value ('iso-cals') were approximately at right angles to each other and equally spaced (except for anthracites). He added subsidiary axes for VM and CV and derived mathematical equations to express the relationship between C, H, VM and CV. He also added lines of equal caking characteristics.

Seyler's classification, chart and relationships together provide the most accurate and useful data obtainable from any classification based upon ultimate analysis for UK coals. UK coals can be grouped together in a band on Seyler's coal chart because they all originated from similar vegetation (tree-ferns etc) at a similar time (Carboniferous) in a similar place (marine bogs in Laurasia). Subsequent depth of burial (pressure) and heat treatment will affect only the extent of coalification (rank) along the band.

Coals from other countries might not lie along the same band as they could have originated from completely different plant types growing in very different environments so will not necessarily have started with the same chemical composition as young UK coals. Coalification will then follow a different path (band).

Shakh Deniz field

Condensate field in the Caspian Sea in which a number of companies including BP, Total and Lukoil have an involvement. Currently producing at about 8 billion m^3 per year of gas, the field will be expanded as further production wells are utilised. Interestingly, the pipeline which will convey the gas to countries including Georgia and Turkey will run close to the **Baku-Tbilisi-Ceyhan Pipeline** so as to eliminate the need to create new space for the pipeline. This of course reduces both pipeline installation costs and environmental impact.

Shale oil, policy of OPEC towards

OPEC countries produce conventional oil from on- and/or offshore reservoirs. A country seeking OPEC membership has to satisfy certain production requirements which are subject to revision by the OPEC Conference, and failure to meet such requirements is likely to result in loss of membership. Might a country be admitted to OPEC on the basis of production of oil from shale? In principle yes, as nothing in the 1960 OPEC Statute precludes this. It has been asserted that the immense shale oil reserves of India are sufficient to make it qualify to become an OPEC country, but there will be many questions to be answered before India's membership of OPEC on the grounds of its shale reserves is at all a realistic proposition.

Shale products, pricing of

When comparing the price of shale oil against that of crude we have to remember that we are not comparing like with like and that some care with terminology is also required. For pricing purposes, 'shale oil' refers to newly retorted shale oil, requiring hydrogenation and desulphurisation before it becomes **syncrude**, on which the Japanese Navy were heavily dependent during World War II. The price of syncrude can reasonably be compared with that of crude oil and referred to the Brent, West Texas Institute or one of the other price indices.

Currently, there is considerable alarm that crude oil prices will rise from $60–65 per **bbl** to a price of $100 per bbl, with all that would entail in terms of world financial stability. Under these circumstances, the belief is that shale oil at $35 to $40 per bbl would be competitive with crude oil. The difference between that and the price of conventional crude would cover the hydrogenation and desulphurisation needed to convert the material as purchased to syncrude. This could be refined to distillates, demanding the same prices as their equivalents from conventional crude. A significant proportion of the cost of producing the shale oil is disposal expenses of the spent shale. However, investment in shale oil production is still under the shadow of **Black Sunday**, which features in discussions on the Internet and elsewhere of large-scale shale oil production. The view taken is that losses would be massive if the price of crude oil stabilised or even fell.

Shale retorting, by-products of

There are two classes of gaseous side product: shale gas, which occurs with the shale and is analogous to **coal bed methane**, and fuel gas, which is the gaseous decomposition product of the kerogen. The latter is so called because it is often used as a fuel at the retort itself. Shale gas is recoverable. If (as is often the case) there has been desulphurisation of the shale oil the sulphur is saleable. The unwanted by-product is of course spent shale and as noted in a previous entry, it is the need to dispose of this which has held back the shale industry from expanding internationally. One should be aware that the processing does not entirely denude the shale of its organic content, and that some solid pyrolysate residue of the kerogen remains in the rock leaving a material rather like a coal 95% in ash. Even this has been burnt in a **fluidised bed** but only on an investigative, not a commercial, basis.

It is possible for there to be shale part way down a well profile for non-associated gas. The best example of such a structure is the **Barnett Shale deposit**.

Shantou, southern China

Site of a major marine terminal for **liquefied petroleum gas** owned by the UK company Fortune Oil. The terminal came into operation in 1998 and has the capability to store, in vessels mounted underground, 60 Mt each of propane and butane for subsequent distribution.

Shea meal

Fruit from the tree species *Vitellaria paradoxa* contains a semi-solid substance marketable in the food industry. Removal of this material in processing is incomplete and retention of part of it by the biomass enhances its reactivity as a fuel which is sold as Shea meal. Shea meal has been exported from Ghana to the UK for use at power stations which co-fire coal with biomass. Its combustion properties are similar to those of **Illipe meal**.

Shell, history of

What we now know as the giant organisation Shell can be traced to early 1830s London when Marcus Samuel went into business selling shells, abstracted from beaches around Europe, to naturalists. The business prospered and in the 1890s Samuel's son and heir diversified into exports of oil for lighting and cooking to the Far East, using such tankers as were available at that time to take consignments of Russian kerosene to Singapore and Bangkok. At the same time there was exploration for oil in the Far East by the company Royal Dutch, and this and Samuel Jr's business merged to form the Royal Dutch/Shell Group in 1907. By about 1912 the group had expanded into North America and Africa. World War I resulted in the loss of some of the capital assets which the Group had built up. A major triumph for the Royal Dutch/ Shell Group was the first non-stop flight across the Atlantic in 1919 by Alcock and Brown: the aircraft in which they made the epic flight was powered by Shell fuel. The company extended from fuels production to petrochemicals in 1929, at about the time that cracking technologies were coming into being.

The period after the World War II saw meteoric expansion of Shell and in the 1970s it had major involvement in the development of the North Sea oil fields. In the early 21st Century, Shell are still very active in the North Sea, in the **Brent field** among others. As the world responded to the environmental consequences of the proliferation of the motorcar, Shell became a world leader in unleaded petrol.

Shell Rotterdam refinery

Europe's largest refinery, with the capacity to refine 418 000 **bbl** of crude oil per day. Gasoline from the refinery is exported to the US: heavier distillate and residual material for use in heating enter the European market.

Shenhua

China's largest coal company. It is currently looking into liquid fuel production from coal, and **Peabody Energy** and **Sasol** are both providing know-how. There are plans both for hydrogenation along the principles of the **Bergius process** and for production of **synthesis gas** which will yield **methanol**.

Sheyang Township

The scene of an explosion involving fluorobenzene (C_6H_5F) in which 22 people died in the Jiangsu Province of China, in July 2006. Halogenated hydrocarbons tend to burn similarly to their unsubstituted counterparts and the halogen goes to the corresponding acid, HF in the case of fluorine. The view has been expressed that the explosion at the Sheyang plant was possibly due to insufficient purging with nitrogen of vessels and pipes prior to admittance of the fluorobenzene.

Shipwrecks, oil leaks from

There have been several such leaks recently. For example, in 2003 a sunken Japanese tanker in Russian waters was found to be releasing oil copiously. The vessel had sunk in 1979. In 2001, the northern California coast experienced oil contamination and birds covered with oil were taken by volunteers to the Wildlife Care Centre near San Francisco. The leak was traced to the vessel SS *Jacob Luckenbach* which sank off California as long ago as 1953 after colliding with another ship. Accordingly, the wreck was accessed and 85 000 gallons of **bunker fuel** removed. That which could not be removed was capped. It is believed that bird life along the coast had in fact been affected for about a decade previously and that thousands of birds had died. The cost of accessing the sunken vessel, removing a portion of the oil and sealing off the rest was almost $US 20 million.

Shizuoka Gas

A Japanese company which supplies gas, previously imported as **liquefied natural gas (LNG)**, to the heavily industrialised Shizuoka Prefecture. Its import terminal is on the shore of Shimizu Bay and is the only such facility in the Prefecture. Japan is a seismically active country and leakage of industrial quantities of LNG would of course add enormously to the consequences of an earthquake. Accordingly Shizuoka Gas have in use LNG-containing storage tanks which will remain intact and retain the LNG during an earthquake of up to 7 on the Richter scale.

Importing about 1 Mt yr^{-1} of LNG, Shizuoka Gas is the sixth largest of Japan's 200+ gas suppliers. The largest is Tokyo Gas, which imports 7 Mt yr^{-1} of LNG. Perhaps because Tokyo itself is within the catchment area of Shizuoka Gas, there has been considerable assistance given by the larger company to the smaller. For example, large tankers of LNG delivering for Tokyo Gas have diverted to Shimizu Bay to deliver part of their payload to Shizuoka Gas, the remainder going on to Tokyo Bay and being delivered to Tokyo Gas. Sale of part shipments of LNG is rare. Nevertheless, the import terminal at Shimizu Bay will accommodate the largest of LNG tankers in anticipation of acceptance of full shipments as the company expands. Also, in 2005 Shizuoka Gas began to purchase LNG from the gas fields off the northwest shelf of Australia in which Tokyo Gas have a significant stake.

Short rotation coppice

Fundamental to the ability of fuel wood to renew itself is rapidity of growth. There is currently much interest and development in the short rotation coppice whereby species such as willow, hazel and poplar are planted for fuel use. Removal of fuel wood from the trees is every 2–5 years depending on the species of tree, and re-growth of the part of the tree from which fuel wood has been removed is not only rapid but sometimes exceeds 100% completion. This leads to a higher yield after a few harvestings than initially. A tree can be so utilised for up to 30 years. A plantation operated on this basis yields fuel wood at a rate of the order of 10 tonnes per hectare per year.

The short rotation coppice is perhaps the most common example of what is currently referred to as **energy crops**. The grass **Miscanthus** is another increasingly important example. Coppiced willow and Miscanthus are the two major energy crops in the UK at present. Naturally, power or **combined heat and power (CHP)** plant using such fuel are most suitably sited in rural areas so that the fuel can be produced locally. One of the largest thermal facilities using short rotation coppice fuel in the UK at present is in Frome, Somerset, where there is a CHP installation producing 7 MW each of power and heat.

Short ton

Unit of weight sometimes applied to coal and equal to 907 kg. A long ton, more commonly known simply as a ton, is 1016 kg. A 'tonne' is 1,000 kg.

Shtokmanovskoye

One of the largest natural gas deposits in the world, located near the Barents Sea in Russia, discovered in 1988. Reserves are estimated at 3000 **bcm** of gas and 27 Mt of oil. It is planned that the field will be put into operation in 2010.

Siemens gas

Producer gas made from coal or **coke** plus air but without any steam. Almost the entire flammable part is carbon dioxide, there being very little in the way of **hydrogen** or hydrocarbons.

Sines

Portuguese coastal location, the site of a major **liquefied petroleum gas** reception facility. Storage prior to distribution of 60 000 Mt is provided for. BP are part owners of the facility.

Singapore Petroleum Company

Founded in 1969 and a major downstream operator in the region having two refineries, one of them on **Jurong Island**. It also has many retail outlets. It is involved in exploration activities with oil companies from other countries in the region, in particular Vietnam and Indonesia.

Sinopec

China's largest oil refining company. Its business is largely the refining of imported crudes. China has been a net importer of oil since 1993, imports having increased sevenfold in as many years. Most of China's domestic oil to date has come from onshore fields in the east and northeast of the country. Chinese demand for oil is expected to reach 296 Mt by 2010 but domestic production is only projected to be 170 Mt by then, meaning that 43% of its oil demand would have to be met by imports. Sinopec are accordingly investing in pipelines to take imported crude to its refineries but are also seeking investment in oil production facilities outside China including those at **Tuban, East Java**

Sipadan

Island the sovereignty of which, and that of neighbouring Ligitan, has been contested by Malaysia and Indonesia. There is believed to be significant offshore oil and gas close to these islands and each country is intending to explore. The situation became confrontational when Indonesia awarded an exploration licence to Total, who duly sent a team. Total's vessel was intercepted by a Malaysian patrol ship.

Sir Baniyas Island

Location of a proposed wind farm for the supply of power to Dubai, UAE. It will be operated by the Dubai Water and Electricity Authority who currently produce electricity thermally at just under 4 GW.

Sirri oil field

Offshore Iranian oil field divided into the Sirri A field and the Sirri E field where production began in 1998 and 1999 respectively. Joint production is currently 120 000 **bbl** per day, there also being **associated natural gas**.

Sixteen tons

Song released on a 78 rpm gramophone record around 1955 by Tennessee Ernie Ford (written by Merle Travis nine years previously), sometimes referred to as 'The battle cry of the American miner'. Its message is that many US coalmine employees in the early 20th century worked and lived in the most appalling circumstances. The refrain to the song goes:

You load sixteen tons, what do you get?
Another day older and deeper in debt
St Peter don't you call me 'cause I can't go
I owe my soul to the company store

This is a reference to situations in which the mining companies had total control over the financial affairs of their employees, who lived in houses owned by the companies for which rent was charged. A miner might also have to pay for hire of the tools he needed to work. Most seriously of all, wages were in scrip, not in currency, and could therefore only be spent at the company store where the miners' families had to purchase their food and household needs. Because of the lack of competition, prices at the company store would often be much higher than elsewhere. Men in such circumstances were therefore usually on a treadmill of debt to the company store.

Sixteen tons demonstrates that it is important to consider the social effects of fuel production and utilisation.

Skye, Isle of
Located off the west coast of Scotland, the island is the site of several proposed wind farms. Already approved is the facility in Edinbain, located in north Skye, which will comprise ten turbines. Each turbine will be 100 m high. Opponents of the idea fear the adverse effects on landscape and tourism.

Slagging and fouling
Slagging refers to the ash deposition processes that take place within a coal-fired power station furnace in the areas that are directly exposed to flame radiation, such as the furnace walls and the widely spaced pendant superheaters.

Fouling refers to the ash deposition processes occurring in areas not directly exposed to flame radiation, such as the more closely spaced tubes in the convective passes of the boiler.

Slane
Device for digging **peat** in such a way that it is obtained in regular pieces. It has two cutting edges, so that an operator can cut a channel in the peat deposit having a width determined by the separation of the surfaces. Peat cut in this way can be lifted out of the ground in a particular thickness. Use of a slane is preferable the alternative method, which involves simply digging it out and then kneading it into regular pieces.

Slate
The list and quantities of products from a refinery.

Sleipner field
Gas field in the Norwegian sector of the North Sea, now depleted. It is intended that it will be used for **carbon sequestration**, that is, carbon dioxide from combustion processes will be disposed of there.

Slurry
(1) Term applied in coal handling practice to fines which have formed a paste-like material owing to water retention after coal washing. It is sometimes beneficial to add some such slurry to a bed of coal in a grate. The reaction

$$C + H_2O \rightarrow CO + H_2$$

provides fuel gases which in effect add the coal volatiles.

(2) Material formed by mixing coal particles, of median diameter $\approx 30\,\mu m$, with water in controlled proportions yielding a stable fluid which can be pumped. Solar drying of such slurry can yield a hard product and this has sometimes been seen as an alternative to making **briquettes** from the coal.

Smuggling of oil
This appears to be rampant in Indonesia, which loses both crude oil and refined products in this way to the extent that annual financial losses are $US 1.4 billion. It is believed that 105 tonnes (≈ 700 **bbl**) of crude oil are smuggled out of Indonesia each month to countries including China and Thailand. There have been over 58 persons arrested in a recent investigation of the matter, 18 of them employees of the state-owned oil company Pertamina. There are other regions of the world where this has occurred, and continues,

Snowy Mountains Hydroelectric Authority
Supplier of electricity for New South Wales, Australia. The Snowy Mountain scheme, which began in the late 1940s, attracted immigrant employees from 30 countries. There are an additional four companies producing electricity thermally in NSW.

SNOX process
One of the most advanced systems for the combined removal of SOx and **NOx**. The SNOX process, developed in Denmark, is located downstream of the particulate control device. The flue gas is reheated and

undergoes **selective catalytic reduction**. It is then further heated and passed to a second catalytic chamber where SO_2 is oxidised to SO_3. The gas is then cooled to condense SO_3 out as sulphuric acid.

SoCal Edison

The largest US purchaser of renewable energy, currently working with Stirling Energy on the construction of a plant for power generation from **photovoltaic cells**. It will occupy an 18 km^2 area in the Arizona desert and power production will be at 500 MW. Construction will actually be by Stirling Energy; the importance of SoCal to the undertaking is that it has made it financially sound by agreeing to buy all of the electricity produced in the first 20 years of operation.

Solar 1

Vessel for oil transportation, which sank in August 2006 in waters close to the Philippines. It had been chartered by Petron, in which the Philippine National Oil Company (PNOC) have a 40 % holding. The vessel had 500 000 gallons of the oil on board. On release, it created an oil slick which affected over 100 miles of the national coastline. The livelihoods of local residents and the tourist industry were heavily affected. Much sea life was destroyed. Outside aid—from Japan and from Indonesia—was sought in the clean-up.

Solid oxide fuel cell

Solid oxide fuel cells use an electrolyte that conducts oxide (O^{2-}) ions from the cathode to the anode. The electrolyte is composed of a solid oxide, usually zirconia (stabilised with other rare earth element oxides like yttrium), and takes the form of a ceramic. Solid oxide fuel cells are built like computer chips through sequential deposition of various layers of material. Common configurations include tubular and flat (planar) designs. Metals such as nickel and cobalt can be used as electrode materials. Solid oxide fuel cells operate at about 1000°C and a pressure of 15 psig (1 barg). Each cell can produce between 0.8 and 1.0 VDC.

The advantages of solid oxide fuel cells are that they:
• support spontaneous internal reforming of hydrocarbon fuels. Since oxide ions, rather than hydrogen ions, travel through the electrolyte, the fuel cells can in principle be used to oxidise any gaseous fuel
• operate equally well using wet or dry fuels
• generate high-grade waste heat
• have fast reaction kinetics

- have very high efficiency
- can operate at higher current densities than molten carbonate fuel cells
- have a solid electrolyte, avoiding problems associated with handling liquids
- can be fabricated in a variety of self-supporting shapes and configurations
- do not need noble metal catalysts.

The disadvantages are that they:

- require the development of suitable materials that have the required conductivity, remain solid at high temperatures, are chemically compatible with other cell components, are dimensionally stable, have high endurance and lend themselves to fabrication
- have a moderate intolerance to sulphur. Solid oxide fuel cells are more tolerant to sulphur compounds than molten carbonate fuel cells, but overall levels must still be limited to 50 ppm. This increased sulphur tolerance makes these fuel cells attractive for heavy fuels. Excess sulphur in the fuel decreases performance.
- do not yet have practical fabrication processes
- the technology is not yet mature.

Solid oxide fuel cells can operate using pure hydrogen or hydrocarbon fuels, just like molten carbonate fuel cells. This results in an inlet fuel stream comprised of hydrogen with or without carbon monoxide.

The reactions at the anode are:

$$H_2 + O^{2-} \rightarrow H_2O + 2e^-$$
$$CO + O^{2-} \rightarrow CO_2 + 2e^-$$

The reaction at the cathode is

$$\tfrac{1}{2}O_2 + 2e^- \rightarrow O^{2-}$$

The O^{2-} ion is drawn through the electrolyte from the cathode to the anode by the reactive attraction of hydrogen and carbon monoxide to oxygen, while electrons are forced through an external circuit from the anode to the cathode. Since the ions move from the cathode to the anode, this is the opposite of most types of fuel cells where the reaction products accumulate at the anode rather than the cathode. Combining the anode and cathode reactions, the overall cell reactions are:

$$H_2 + \tfrac{1}{2}O_2 \rightarrow H_2O$$
$$CO + \tfrac{1}{2}O_2 \rightarrow CO_2$$

The fuel cell therefore produces water, regardless of fuel, and carbon dioxide if using a hydrocarbon fuel. Both product water and carbon dioxide must be continually removed from the cathode to facilitate further reaction.

Solid polymer fuel cell
See **polymer electrolyte membrane fuel cell**

Solid waste/crude oil equivalence
In discussions of the use of solid wastes such as **municipal solid waste (MSW)** as fuel, there is often the assertion that a tonne of such waste burnt is a **barrel** of crude oil saved. This is examined below.

> 1 tonne MSW, calorific value ≈ 7 MJ kg^{-1}. The heat released on burning
>
> $$= 7 \times 10^6 \times 10^3 = 7 \text{ GJ}$$
>
> 1 barrel of crude has volume 159 litre $= 159 \times 10^{-3}$ m^3
>
> If density is 950 kg m^{-3}, the mass of one barrel $= 151$ kg
>
> Letting the calorific value be 44 MJ kg^{-1}, the heat released on burning
>
> $$= 44 \times 10^6 \times 151 \text{ J} = 6.6 \text{ GJ}$$

Agreement is evident, there being some variability in the calorific value of the waste and the density of crude oil. It is therefore an appealing rule of thumb that a tonne of MSW is equivalent to a barrel of oil.

Solomon Islands, electricity supply in
The Solomons form part of the Pacific region known as Melanesia. Being a major sugar producer **bagasse** is available as fuel, and there is also timber in abundance. Notwithstanding these resources the Solomons currently generate power at a total rate of about 3.5 MW, entirely with imported oil. There are projects planned to press the bagasse and wood fuels into service in order to eliminate the dependence on imports. The imports themselves have been threatened in recent years by non-payment of the suppliers and attacks on tankers.

Songo Songo gas fields
Songo Songo is an island off the east coast of Africa. Wells for gas production have been drilled on the island and in the nearby waters and, after processing on the island, this is sent by pipeline to the Tan-

zanian capital Dar es Salaam where its primary use is power generation. Tanzania did not develop as hoped during the early years of its
independence: future development will require a reliable electricity
supply. Production of gas is currently just under 1 million m³ per day
and electricity generation will be by gas turbines, not steam turbines.

Sontara®
Industrial wipe material patented by E I du Pont de Nemours. Once
so contaminated as to be unfit for further use, the material is suitable
for fuel use.

Soot
Formed from the incomplete combustion of coal, oil, wood and other
carbonaceous materials. Soot from diesel vehicles, particularly visible
to observers on start-up or during a gear change, is toxic to humans
and a source of global warming.

Soot blowing
A technique used in coal-fired power stations to remove slag deposits
from the boiler walls. Lances are periodically injected into the boiler,
through which jets of high-pressure air are aimed at the deposits. No
soot is involved in the technique, despite the name.

Sorbent injection
See **desulphurisation**

Sorghum
Cereal crop originating in Africa and naturalised to the US in the
early 1800s. It is an alternative to corn in the production of **ethanol** as
fuel, notably in Kansas and Minnesota.

Sour gas
Natural gas having an unusually high level of hydrogen sulphide which,
on combustion of the gas, will convert to **sulphur dioxide**. Its use in facilities such as power stations is not necessarily precluded by its sourness. A
power station might use up most of its sulphur credits over a short period
by burning such gas. The disadvantage of loss of sulphur credits will be
offset by the fact that the gas, being sour, was cheap in the first place. Such
usage of sour gas is however the exception rather than the rule. Sour gas
obtained at an offshore platform which produces both oil and gas, for
example, will usually be diverted to the platform **flare** for destruction.

Source rock

A sedimentary rock sequence containing organic-rich material capable of producing and expelling petroleum under favourable temperature and pressure conditions. Source rock must contain a certain amount of organic matter. The burial history will determine whether the source rock will yield and expel petroleum as oil, gas or a mixture of both products on maturation. Proto-petroleum is released from source rocks to reservoirs.

South Africa, issue of oil and gas exploration licences by

In post-Apartheid South Africa there have only been two rounds of licences, one in 1994 and one in 2005. South Africa imports oil and also obtains coal-derived liquid fuel from **Sasol**. Proposed exploration in the near future will be in the Tugela Basin off the coast of Kwazulu-Natal at water depths of about 500 m.

South Arne field

Oil field in the Danish Sector of the North Sea where production began in 1999. It is estimated to contain over 100 million **bbl** of crude oil and currently produces about 40 000 bbl per day. Amerada Hess have a greater than 50% holding in the field.

South East London combined heat and power (SELCHP)

A pioneering organisation which takes wastes and charges a fee comparable to a **merchant incinerator** for its treatment. The waste is then burnt to raise steam for a **combined heat and power** cycle. The power produced can be sold to the grid and the heat produced sold to local residents for domestic heating.

South Korea, solar power in

The largest solar power plant in South Korea is in Seoul and has an output of 1.2 MW. South Korea is seen as a promising part of the world for solar power, and this could relieve the country's current heavy dependence on imported fossil fuels. Accordingly the government is offering a guaranteed price per unit of solar electricity to investors who participate in solar power production.

South Pars field

An extension into Iranian waters of the **Qatar North Field**, having just under half as much gas as the Qatar North Field itself. It supplies 60% of Iran's own natural gas requirements, and a significant amount for export.

Southampton, geothermal power at

Southampton is the location of the only **geothermal electricity** generation in the UK at the present time, at about 2 MW level. Though the UK is not rich in geothermal resources, it is believed that it could contribute to 2–3% of the country's power needs, although the necessary development work would be prohibitively costly.

Soybean

A source of **biodiesel,** soybean oil is currently the most expensive of the liquids used but has some advantages over its competitors. One is that soybean-derived biodiesel is particularly suitable for use in colder regions.

Spalling, of concrete structures at offshore installations

If, as with **Brent Bravo**, the support for an offshore production platform is concrete, spalling of the surface of the concrete above the waterline is sometimes observed after a long period of service. When such spalling was discovered inside the utilities leg at Brent Bravo, an investigation concluded that there was no loss of support capability of the concrete. It was noted, however, that there was danger from possible detachment and descent of concrete debris. Action to prevent this was taken by containment of the affected concrete.

Spindletop field

The site of an initially uncontrollable release of crude oil in Beaumont, Texas, in response to exploratory drilling in January 1901. The event is considered to mark the beginning of the oil industry in the Gulf States. There had been drilling at Spindletop Hill for a number of years previously by the Gladys City Oil, Gas, and Manufacturing Company. Advice from one of the leading petroleum geologists at the time was that there was oil there in abundance, a conclusion largely due to the presence of **salt domes**. Seepage of oil at the Spindletop field had in fact been observed some time before the dramatic gush. The release was not capped for several days and the local area was deluged in crude oil.

The Spindletop field exceeded the combined known oil reserves of the rest of the world at that time and many companies familiar to us today resulted from its development. One was the J M Guffey Oil Company, which we now know as Gulf Oil. Another was Magnolia Oil Company, which later changed its name to Mobil. Yet another was the Texas Oil Company, later Texaco.

Sriracha

Coastal region of Thailand south of Bangkok, of interest in fuel technology for two reasons. The Sriracha refinery, operated by Esso, began operations on a fairly small scale in 1967. Expansion has raised the capacity of the refinery from 7000 **bbl** per day to 170 000 bbl per day.

Secondly, there was a major spillage of diesel into the sea when a tanker bearing diesel collided with a cargo boat in March 1994.

Spirit of coal

Also known as coal gas, the name given to gases produced from the partial oxidation of coal to form **coke**. Formerly used as a fuel and for illumination.

St Croix

Location in the US Virgin Islands about 550 miles from the Venezuelan coast. It is the scene of a major refinery jointly operated by Amerada Hess and Petroleos de Venezuela with a capacity of 495 000 **bbl** per day. It receives crude largely though not exclusively from Venezuela; some crudes from West Africa having been carried by tanker about 4500 miles are refined there. Amerada Hess have another major refinery at Port Reading, NJ. Products from the two refineries are sold along the entire east coast of the US, from Florida to New York.

St Lawrence River

One of the world's major sources of **hydroelectricity**, there being plants drawing on the river both in Ontario and in the state of New York.

Staged combustion

Coal combustion using coal in pulverised form, such that only a proportion of the necessary air is initially admitted, the remainder being admitted after some combustion has taken place. Broadly speaking the volatiles burn in the initially admitted air and the residual solid burns in the air admitted at the second stage. For all ranks of coal this has been shown to lead to significant reduction in the formation of **thermal NO_x**.

Standard Oil

Founded by J D Rockefeller in 1870 and the first major commercial enterprise in hydrocarbons. It later expanded into other states including California, Kentucky and Iowa. Standard Oil in the respective states became independent companies in 1911, when under the antitrust laws the nationwide Standard Oil was broken up.

By 1941, Standard Oil began marketing under the brand name Esso (the phonetic pronunciation of an acronym of Standard Oil) within certain states. In 1972, Standard Oil of New Jersey renamed itself Exxon.

Stat

Name of the first tanker to transport oil across the Atlantic, in 1879. Previously cargo vessels had been used and the oil distributed among a large number of barrels.

Staten Island, NY

The site of significant hydrocarbon activity, which led to major litigation in 2001. There were claims that in the early 1990s at the Mobil terminal on Staten Island there had been the release of 'benzene tainted wastes' without a hazardous wastes permit. The successor company ExxonMobil were fined $8.2 million and required to pay $3 million for restoration of affected parts of Staten Island. A statement made by the company assured that payment was not an admission of guilt, but a way of avoiding lengthy legal proceedings.

Statoil

Norwegian oil concern. One of its current activities is **carbon sequestration** at the **Sleipner field** in the North Sea. Statoil have also had an interest in **alpha olefins** production.

Steam, properties of

Water in gaseous form or in vapour form if it is in phase equilibrium with liquid. It is the working substance in many thermodynamic cycles including the **Rankine cycle** which is the basis of most thermal generation of electricity. Steam can be saturated, in which case it is in equilibrium with vapour, or superheated in which case there is no vapour, the entire mass being in the gas phase. At any one temperature, steam and liquid water can co-exist under equilibrium conditions at only one pressure, determined by nature. This follows from the phase rule. For example, water and its vapour can be in equilibrium at 100°C (373 K) only at 1 bar pressure. Since 1 bar corresponds to atmospheric pressure, 100°C is the normal boiling point of water. Tables of pressure and temperature for saturated steam are widely available. Knowledge of the pressure and temperature of a particular sample of steam gives no indication at all of what proportion of the total mass is in the vapour phase. This proportion is referred to as the dryness fraction and is important in thermodynamic applications. There are means of measuring it *in situ* with a throttle calorimeter.

Superheated steam has, in the terminology of the phase rule, two degrees of freedom, therefore superheated steam requires knowledge of both pressure and temperature in order to be fully specified. If the pressure and temperature are both known, the other properties, including the specific **enthalpy**, are fixed. The properties of superheated steam at specified temperature and pressure, such as the specific enthalpy and the specific **entropy**, are available from steam tables.

Steam coal
The name given to coal that is suitable for raising steam in a **coal-fired power station**. Formerly used in railway locomotives and ships.

Steam cycle in the coal-fired power station
Coal fired power stations generate electricity via steam. Steam at high temperature and pressure is generated in the power station boilers, driving the **turbines** to generate the electricity. By utilising steam as the generating medium, only about one-third of the available energy in the coal is utilised. The remainder is wasted by passing to the atmosphere via cooling towers or water cooling.

The basic **Rankine cycle** is not suitable for practical application in a power station due to the low efficiency of power generation and to operational limitations. Modern power stations are based on a modified Rankine cycle which includes reheat to raise the cycle efficiency.

Subcritical boilers typically take the form of drum boilers in which steam and water from the evaporative part of the cycle (the furnace walls) are separated in the drum. Water from the drum passes down the downcomer and enters the furnace at the base, passing up the water walls and turned into steam before returning to the drum. This wet steam is not of high enough quality to pass to the turbine directly, so is first passed back through the furnace superheater (large pendants of tubes hanging in the top of the furnace). The superheated steam passes to the high-pressure turbine where energy is extracted during expansion. As some energy remains in the steam, it is passed back to the furnace and passes through the reheater (pendants situated next to the superheater) and back to the intermediate pressure turbine. The expanded steam passes, via a crossover, to the low pressure turbine where any remaining useful energy can be extracted. The exhausted steam is passed through a condenser, de-aerated and sent back to the steam drum, via the **economiser**, for the cycle to begin again. The cycle is closed, with purified water being added to top up the system as necessary.

The circulating water cycle in the power station is more visible. Water from a local source, typically a river, is pumped through the low-pressure turbine condenser to cool the exhaust steam. This low-grade steam passes to the cooling towers where it condenses and is either returned to the source or recycled with a top-up from the source.

Coal fired power stations built over the last 35 years and using subcritical steam conditions have net efficiencies of 36–38%. At 38% efficiency, a 600 MW generating unit will require 1580 MW of thermal energy. This is equivalent to 62 kg s^{-1} of coal with a **calorific value** of 25.51 MJ kg^{-1} (6100 kcal kg^{-1} net as received). At 38% efficiency, a 1000 MW$_e$ station burning bituminous coal at an overall load factor of 70% will consume 2.28 Mt yr^{-1}.

Subcritical steam cycles operate well below the steam/water critical pressure of 221.2 bar. Plant thermal efficiency and, therefore, environmental performance are enhanced by increasing the operating temperature and pressure. Supercritical-pressure boilers cannot use this drum design, with either forced or natural re-circulation of water, because there is no distinct water/steam phase transition above the critical pressure. They must, instead, be of a once-through design. The form of the steam turbine employed under supercritical conditions is the same as that for subcritical cycles although the temperatures throughout the system are all higher. The main impact of the supercritical cycle is to increase the overall plant thermal efficiency, thereby reducing the fuel consumption per unit of electricity generated. As well as reducing fuel costs, this is a highly practical way of reducing CO_2 emissions. Replacing a typical subcritical plant of 38% efficiency with a state-of-the-art advanced supercritical design of » 45% efficiency results in ≈15% reduction in CO_2 emitted per kWh electricity generated. Similarly, all other emissions (**NOx** and SOx) are reduced pro rata with the fuel consumption.

Steam reforming

The reaction of hydrocarbons with steam to form a gaseous mixture comprising oxides of carbon and nitrogen. This might be used as a fuel gas or as **synthesis gas**. Low-value refinery products or **cracking** residues are often suitable feedstocks.

Steam turbine
See **turbine**

Stevenson Expressway, Chicago
Named after Governor of Illinois Adlai E. Stevenson, and the scene of a major **ethanol** leak from a road tanker in October 2005. When the tanker overturned and began leaking liquid the State Police were initially uncertain of its identity, believing that it might be gasoline. All lanes of the Expressway were closed for a number of hours.

Straight vegetable oil (SVO), vehicular use of
Biodiesel has to be carefully specified to be suitable for blending with mineral diesel or as a replacement fuel for an engine designed and built to run on diesel. SVO is vegetable oil sold for culinary purposes requiring no stringent **viscosity** specification. Vehicles can be adapted to run on this sort of oil in unused or in waste form, although in the latter case filtration is necessary. The adaptations to make a diesel vehicle run on SVO are fairly major, requiring installation of a **greasecar system**.

Stranded gas
Natural gas occurring offshore distant from established fields such as those of the North Sea or the Gulf of Mexico. Gas from such fields can be taken to a **barge** for collection and processing. At present the emphasis is on the manufacture of **gas-to-liquid fuels** from gas obtained at stranded fields.

Strategic petroleum reserve (SPR)
Crude oil held in **salt domes** along the Gulf Coast for use in a contingency, set up during the presidency of Gerald Ford. Authority for the release of oil from the Reserve can be granted only by the President of the United States. Late in 2006 the Reserve stood at 688 million **bbl** (out of a capacity of 727 million bbl), and oil in the reserve is by no means all US in origin. The oil currently in the reserve was bought at an average price of \$27.25 per bbl.

The most recent withdrawal was in response to **Hurricane Katrina**, when a withdrawal of 20.8 million **bbl** was made. This enabled refineries in the Gulf States to resume operations and at least partially restore gasoline availability nationally, which were strongly affected by the Hurricane.

One-quarter of the domestic oil of the US is from the Gulf of Mexico. The **Northeast heating oil reserve**, which has been in existence since 2001, is considered as part of the Strategic Petroleum Reserve. In December 2006, the DOE announced it had identified the salt domes of Richton, Mississippi, as an additional site at which to expand the SPR (as well as at three other sites) to give a total of one billion barrels capacity.

Straw, fuel use of
There is fuel use of straw in parts of the world including Denmark (≈ 3 Mt yr^{-1}) and Croatia (≈ 0.8 Mt yr^{-1}). Its **calorific value** is ≈ 15 MJ kg^{-1}. At **Ely Power Station** in the UK there is significant electricity generation with cereal straw as fuel.

Stybarrow
Oil field off northwest Australia to be developed by BHP. It is expected that production will begin in 2008. Its contents are estimated to be in the range 60–90 million **bbl**.

Sub-bituminous coal
Coal of rank between **brown coal** and **bituminous coal**, having a carbon content of around 60%. Sub-bituminous coals are black in colour but tend to lack the gloss that some bituminous coals have. When carbonised, sub-bituminous coals yield a solid residue which is not fused, and so they are unsuitable for coke manufacture although they have been used to make **retort coal gas**.

Compilations of statistics and standard texts on fuel production unfortunately tend to lump lignite and sub-bituminous coal together which is unfortunate as the two are different in their properties and not in general interchangeable in utilisation. One major difference is that lignite, as mined, is usually 50+% in moisture, whereas sub-bituminous coals have a moisture content of about half that. Where sub-bituminous coal is used in place of a lignite in power generation the effect will, other things being equal, be an increased power output. In the US there are abundant deposits of sub-bituminous coal in Colorado, Wyoming and Montana. The **Black Thunder mine** in Wyoming, which is composed of sub-bituminous coal, is one N America's most productive mines having yielded 1000 Mt since operations there began in 1977. The mine at **Gillette, WY**, began operations much longer ago and is also very productive of sub-bituminous coal. (The place where the first coal production in Wyoming took place in the 1860s is now known as Carbon County.) One of Europe's major producers of sub-bituminous coal is the Czech Republic. Most coals mined in Indonesia are sub-bituminous.

Sub-sea oil and gas, temperature of

A widely applied rule is that oil and gas *in situ* at an offshore field rise in temperature by 3°C for every 100 m depth. On this basis, oil at a depth of 500 m (approximately the depth at which oil is obtained from the **Franklin and Elgin fields** in the North Sea) will be at a temperature of 170°C.

Sub-sea pipeline, corrosion protection of

There are coatings of various kinds which can be applied to the outside or (less commonly) the inside of a pipeline to provide resistance to attack from salt water. The principles of cathodic protection, first applied by Sir Humphry Davy to ships in the British Navy, are at the present time extensively applied to pipelines.

Corrosion of the steel of which a pipeline is composed is an oxidation process: the iron in the steel loses electrons. In cathodic protection the pipeline is in electrical continuity with a metal such as zinc which loses electrons more readily than iron does. Thus an electrochemical cell is set up in which the zinc is the anode and the pipeline steel the cathode. Electrons transfer from anode to cathode, maintaining the condition of the latter to the eventual total loss of the former hence the term 'sacrificial anode'. The performance of cathodic protection can be monitored *in situ* by measuring emfs at mV level in the pipe metal using an instrument mounted on a remotely operated vehicle (ROV).

Sub-sea pipeline, longest in the world

The joint Norwegian-German company HydroWingas is currently constructing a gas pipeline, due to be completed in 2007, to carry natural gas from the **Ormen Lange** field to the northeast coast of England. It will be 1200 km in length, making it the longest single sub-sea pipeline in the world, and will provide about 20% of the UK's natural gas needs. UK self-sufficiency in natural gas ceased in 2005, and this pipeline will help meet the deficit.

Sudan, oil production in

The 2001 figure of 262 million **bbl** for the oil reserves of Sudan had risen to 563 million bbl by late 2005, as a result of exploration in the intervening years. Exploration continues by companies including Total and Marathon Oil. Production of oil in Sudan was very low until 1999 when the export pipeline to Port Sudan was completed. There have been social and political problems in the development of Sudan's oil assets, not least allegations by human rights groups that people have been forcibly uprooted from their homes to enable the oil fields to be expanded. There has also been sabotage of oil production facilities. Production of crude oil in Sudan in 2005 was 500 000 bbl per day. The country has a number of refineries, the most recently commissioned being the Khartoum refinery which came on line in 2000 and has a capacity of 70 000 bbl per day. There are three smaller refineries, including that at Port Sudan. Some of the refined material from the Khartoum refinery is piped to Port Sudan for export. Sudan itself imports only jet fuel, being self-sufficient in all other petroleum products. Sudan has recently been invited to join **OPEC**, as has Angola.

Suezmax
The largest size of fully-laden oil tanker that can safely navigate the Suez Canal. The canal has no locks but the maximum depth below water for shipping is 16 m. This is too shallow for most supertankers.

Suffocation, deaths due to in hydrocarbon accidents
Sometimes when there is a rapid leak of methane, death occurs by lack of oxygen and not due to the flammable nature of the gas. This happened at the 2003 **Brent Bravo** accident. In mining accidents this can also happen, and has to be distinguished from deaths due to inhalation of **afterdamp** which, unlike methane, is toxic.

Sullom Voe
The location of an oil terminal in the Shetland Islands (off the northeast coast of Scotland). Construction on the oil terminal began in 1975 and it was finally completed in 1982. At that time, it was the largest oil terminal in Europe. It receives crude oil by pipeline from North Sea oil fields for loading on to tankers.

Sulphite liquor
A material similar in origin and use to **black liquor**. Some processes for the conversion of wood to paper involve treatment of the wood in the form of chips with water, **sulphur dioxide** and ammonia at temperatures around 150°C. The residue after pulp removal is called sulphite liquor, and like black liquor, is suitable for fuel use.

The US pulp and paper industry produces ≈3 Mt of this material annually. In addition to fuel use, there is some interest in the production of fermentable organic compounds from it. One approach to this involves breaking down the lignite with ozone.

Sulphur dioxide

A chemical compound with the formula SO_2, in the form of a gas at ordinary temperatures. Most fuels contain measurable amounts of sulphur and it is the fate of *all* fuel sulphur to go to sulphur dioxide on combustion even if conditions are not powerfully oxidising, that is, if the combustion is fuel-rich. Sulphur dioxide is harmful to the respiratory tract and it also affects vegetation and causes acid rain. **Emission standards** and **ambient standards** apply. When the atmospheric sulphur dioxide concentration in a city is monitored, an anomaly might well be reflected by a rise in deaths from asthma and bronchitis. The only possible exception to the rule that all fuel sulphur is converted to sulphur dioxide is that when hot post-combustion gas undergoes a turbine cycle, sulphur trioxide (SO_3) can be present in significant amounts. The reason for this is that high temperatures favour formation of the trioxide which will not be formed if the post-combustion gases are rapidly allowed to cool, as they are in most applications. In a gas turbine, however, elevated temperatures are sustained during the cycle, enabling SO_3 to be formed. If excess air was used in the combustion there will be several per cent of molecular oxygen in the gas entering the turbine, promoting the formation of SO_3.

Sulphuric acid

Inorganic reagent with chemical formula H_2SO_4, used for purification of distillates including gasoline. It removes some of the sulphur from a distillate and also gum-forming constituents. In refining, a small degree of polymerisation can sometimes occur, giving a distillate an unattractive brown hue. Sulphuric acid breaks down such polymer structures and so improves the colour. In expressing the concentration of sulphuric acid for use in a hydrocarbon purifying operation, the **Baumé scale** is used.

Sumatra, marginal oil and gas fields in

The Indonesian state-owned oil company Pertamina is inviting collaboration from foreign operators in the development of over 40 oil and gas fields with relatively small potential (marginal fields) in southern Sumatra. **Sinopec**, who already have a share in the refinery at **Tuban, East Java**, have responded favourably. Marginal fields become more viable when oil prices are high and when at least part of the necessary infrastructure is already in place. The latter condition is fulfilled in the case of the fields in southern Sumatra. Hydrocarbon from wells only 50 km out to sea is brought ashore in southeast Sumatra. There is also offshore activity off the north coast of Java and the platforms there are networked with those off the Sumatra coast. Bunker fuel from Sumatra was used to power the Japanese aircraft carriers at Pearl Harbour.

Summerland oil field
Located in Santa Barbara, County California, first yielding oil in 1895. About forty wells had been drilled on the beach, and it was observed that the wells closest to the sea were the most productive. Piers extending a short distance over the sea were constructed and sub-sea wells drilled from them. The area containing those wells was formally named Summerland Oil Field (Offshore Area). By the turn of the century there were hundreds of wells in the Offshore Area although management and co-ordination were poor and by 1920 only a few were still producing. The term 'offshore oil field' therefore entered the vocabulary of the industry in 1890s California.

Sunoco
(1) Canadian oil company, based in Ontario with a refinery at Sarnia, which began operations in 1953 and now has a throughput of 70 000 **bbl** of crude oil per day. It is totally independent of the US company of the same name which is described below, with a very similar logo. See http://www.sunoco.com/ to distinguish the two companies.

(2) Philadelphia based petroleum and **petrochemical** company having been in business since 1886. It operates five refineries in the US and has about 4500 gas stations in the eastern US. It also has 4500 miles of crude oil pipeline. There was leakage of oil from of one of the company's pipelines in **Tinicum Marsh**, **PA**, in 2000.

Supermajors
The collective noun given to the world's three largest oil companies: BP, ExxonMobil and Shell.

Supermarkets, sale of gasoline by
In the UK at the time of going to press 30% of retail gasoline sales are through supermarkets including ASDA and TESCO, yet only 10% of forecourts are at supermarkets. This clearly reflects the very competitive prices at which gasoline is sold at supermarkets. The supermarket companies are not in the refining business but buy gasoline from the major oil companies and are at an advantage in so doing in two respects. First, there is the obvious buying power of the supermarket chains and their strong position to negotiate prices. Secondly, particu-

lar supermarkets obtain petrol from local refineries irrespective of which oil companies run them and in so doing cut down on transportation costs to the forecourts.

> In February-March 2007 there were difficulties with petrol purchased at certain supermarkets in the UK; cars having been filled with such petrol kept cutting out. It is believed that silicon, probably in the form of silicone oil or silicone grease, had found its way into the petrol. On combustion of the petrol this would have been converted to silica particles, which clogged the oxygen sensor at the exhaust causing it to fail in its role in engine management.

Supertanker
An oil tanker of the largest size category having a payload of millions of **bbl**. Such vessels are too large to enter some of the oil-receiving ports of the world so the oil has to be transferred offshore to smaller vessels. Supertankers cannot pass through the Suez canal (see **Suezmax** and **Panamax**) and some are even too large to safely enter the English channel. **Exxon Valdez** was a supertanker.

Su-pa-tanka
Term used in Japan for a **supertanker**, a simple phonetic reproduction of the English word. Some of the earliest supertankers were commissioned to meet the sharply rising demand for crude oil imports to Japan following completion of the **Tokuyama Refinery**, and also that at Hokkaido. The present oil tanker fleet of Japan can ship 80% of the total imports.

Sustainability
Sustainability is a concept relating to the continuity of social, economic, institutional and environmental aspects of human society, as well as the non-human environment. It is intended to be a means of configuring activity so that society is able to meet its needs and express its greatest potential in the present, while preserving biodiversity and natural ecosystems, and planning and acting for the ability to maintain these ideals in a very long term.

Sustainable city
A more sustainable city has fewer inputs of energy, water, food etc., and fewer waste products such as heat, air pollution, water pollution

and **municipal solid waste** than a less sustainable city. Cities can be made more sustainable by means of:

- green roofs
- green transport
- sustainable urban drainage systems
- energy conservation
- better garden and landscape design for water conservation.

Sustainable energy
Energy sources which are not expected to be depleted in a timeframe relevant to the human race, and so contribute to the sustainability of all species. Sustainable energy sources are most often regarded as including all renewable sources, such as solar power, **wind power**, wave power, geothermal power, tidal power, and others.

Swamp forests
Present day scenes of vegetation growth in fresh water, which are believed to reproduce the conditions prevailing in the coal age. One of the best-known examples is in Sumatra; it is about 300 square miles in area and has the appearance of logs and other plant debris immersed in brown stagnant water. Another example is in North Carolina, US, and is called 'Dismal Swamp'.

Sweet gas
Hydrocarbon fuel gas free from sulphur compounds.

Switchgrass
Energy crop grown and used in the US Midwest. It is also being used in a scientific study in which its photosynthesis rate at midday is compared with that of energy crops of the genus **Miscanthus**. The Miscanthus crops photosynthesise 25–30% more rapidly than the switchgrass. Switchgrass, unlike Miscanthus, is indigenous to North America.

Switzerland, energy scene in
This country imports all of its coal, oil and gas. Having no domestic oil at all it was particularly vulnerable during the **OPEC** export restrictions in the early 1970s. This came as a wake-up call, and one response was the commissioning of two further nuclear electricity plants in Switzerland in addition to the previous three. At present **hydroelectricity** generation provides 61% of Switzerland's power, 35% being generated at nuclear power stations and the balance being at-

tributable to small-scale generation including that at a handful of **wind farms**. There is investigation of geothermal sources which have previously been regarded as too deep for access.

Switzerland imports about 13 Mt of crude oil per year, chiefly from Nigeria and from Libya. There is also import of material already refined, from Germany, France, Italy and the Netherlands. Natural gas is imported chiefly from Germany, Russia and Holland.

Syncrude

(1) Term for shale oil at a stage where it is ready for refining having been retorted, hydrogenated and desulphurised. In the refining, cuts are taken to give the equivalents of gasoline, naphtha, kerosene and so on.

(2) A **heavy crude** or fuel from **bitumen** having been modified, by hydrogenation or by blending with light material, to bring it to within the **API gravity** of conventional crude oil in readiness for refining.

Synthesis gas

Generic name for gas made from solid or liquid feedstock and subsequently processed, possibly to give a simple organic compound such as formaldehyde or **methanol**. Before the petrochemical industry came into being this was the major route to industrial organic chemicals. Nowadays, synthesis gas is often further reacted to form petrol substitute or extender.

Synthetic motor oils

These lubricant oils are synthesised by blending organic compounds in controlled amounts. **Alpha olefins** of C_{20+} form the base, and other organic compounds including diesters are blended to give synthetic oils of differing specifications. In contrasting synthetic motor oils with conventional ones using a **petroleum** residue as base, we must remember that the alpha olefins were made from ethylene having been made by cracking petroleum material. The other ingredients are also likely to have originated from petroleum. The point being made is that so-called synthetic motor oils could not be produced in a hypothetical petroleum-free milieu.

S Zorb™

Gasoline sulphur removal agent developed by ConocoPhillips. An S Zorb™ installation currently in use in Beijing handles 30 000 **bbl** per day of gasoline with regeneration of removal agent. ConocoPhillips refineries in the US using S Zorb™ include that at **Wood River, IL.**

T

Tabiyeh liquefied petroleum gas (LPG) rail facility, Syria

Scene of a major terminal and handling facility for **LPG**. The LPG originates from heavy components of **associated gas** at the nearby Deir Ez Zor oil field, some of which is re-injected into the wells: some is pipelined and used as fuel by local industries. The balance is converted to LPG and sent by rail to the terminal facility for distribution.

Tacky cloth

A form of fibre glass, suitable for making temperature-resistant gaskets for installation in steam raising plant, an alternative in such an application to **non-asbestos materials**. The performance and reliability of gaskets or other devices made from tacky cloth can be improved if, before addition of resin, the glass fibres undergo the process known in the textile industry as carding. This involves cleaning by use of a toothed device.

Tacoa, Venezuela

Scene of a fuel oil explosion at a power station, December 1982. Two employees were killed in the initial explosion at one of the fuel oil tanks. The fire service declared it safe to allow the tank of oil to burn itself to depletion. A crowd of employees as well as members of the public gathered to watch and over 150 of them died when a fireball occurred eight hours after the initial explosion.

Tahe field

Newly discovered oil and gas field in Xinjiang, northwest China, estimated to contain 8 billion **bbl** of oil and 50 **bcm** of natural gas. **Sinopec** are developing the field, and investment in it will counterbalance that in new pipelines for imported crude oil which has featured strongly in the company's recent portfolio.

Taiwan, oil tanker scrapping

Taiwan has for the last 20 years been the destination of many oil tankers having ceased to be serviceable. Ferrous metal from the tanker so broken up can be encapsulated by concrete and used in the building industry. Steel producers themselves seldom buy back such scrap. The entire ship-breaking industry—tankers and other vessel types—provides not more than 3% of the scrap ferrous metal on the world market. A tanker might be berthed at **Johor** for a period pending a final decision to scrap.

Taiwan, reclaimed land as sites for power and petroleum plants in

There has been reclamation of a 2096 ha area of land in Taiwan, much of which is or will soon be occupied by plant that will assist in meeting the country's energy needs. There will be a thermal power plant with three steam turbines each of 600 MW. All of the generated power will be transferred to the national grid. There will also be a crude oil refinery with a capacity of 450 000 **bbl** per day and a naphtha cracking plant to provide feedstock for the petroleum industry.

In the reclamation process, over 6 million m³ of concrete were used and 4.5 million m of pile.

Tak Province, northern Thailand

The scene of a huge shale deposit, where exploratory drilling first took place as long ago as 1935. Proven recoverable reserves are estimated as 810 Mt. There is no activity there at the present time.

Tallow

Animal fat once used to make candles. It is still important in 21st century fuel technology, as liquid waste comprising tallow and used vegetable-based cooking oils are being refined into blendstock for motor fuels. Fuels for compression-ignition engines containing up to 5% of such material, balance diesel, are currently available in places including Scotland. There is total interchangeability with fuels comprising 100% diesel.

Tantawan field
Oil field in Thai waters, worked by **Chevron**. Current output 10 000 **bbl** per day.

Tar
A highly viscous liquid product obtained from the distillation or thermal decomposition of carbonaceous materials such as coal and wood. Tars contain compounds such as phenol, cresols (methylphenols) and xylenols (dimethylphenols) with industrial uses. Natural bitumen is also known as tar.

Phenol   *m*-cresol   2,6-dimethylphenol

Target depth in offshore drilling
The **Roncador field** represents the deepest drilling (about 5500 ft) yet achieved for offshore oil and gas production. The projection of the industry is that drilling at 3000 m (9850 ft) will have been accomplished not later than 2015, perhaps by 2010. In all blueprints for drilling at this extreme depth, a **floating production, storage and offloading (FPSO)** features.

Tasmania, wind power in
This island state off Australia raises an impressive 5% of its electricity in this way. There is one **wind farm** in Tasmania currently operational and one approved for building. The newer wind farm will have 25 3 MW turbines.

TDF
See **tyre derived fuel**

Teak
A type of wood, used in the manufacture of furniture and ornaments. The world's largest producer of teak is Thailand, where the waste from manufacturing is diverted to fuel use.

Tehachapi, CA
Located 72 miles from LA, the site of a **wind farm** with around 5000

wind turbines, the second largest in the world. Twelve private companies are represented, and together produce about 800 MW of electricity.

Teikoku Oil

Japanese oil and gas exploration company currently at work in the **East China Sea**.

Tempered liquefied petroleum gas

Liquefied petroleum gas (LPG) diluted with air by about a factor of 3.5 to give a gas of **calorific value** approximately 25 MJ m^{-3} which can be used on a burner designed and adjusted for natural gas. The quite severe dilution is necessary to reduce the **thermal delivery** and also the flame speed which with undiluted LPG is much higher than with natural gas. The proportions of air and LPG in tempered LPG are however such that the mixture is well on the high side of the upper flammability limit and so will not burn without further air. It is clear that interchangeability of tempered LPG with natural gas is brought about by control of **excess air**, some of which enters the burner not as atmospheric air but as a fuel diluent.

Tennessee Valley Authority (TVA)

One of the earlier developers of hydroelectric power, and now one of the largest suppliers of electricity in North America. Its creation can be attributed to President F D Roosevelt in 1933 when the construction work required had the advantage of getting unemployed men into jobs. Power generation was initially entirely by **hydroelectricity** but the authority later diversified: coal-fired and nuclear, the latter including **Brown's Ferry**. By the beginning of the 21st century, over 60% of the power provided by the TVA was from coal. Nevertheless, hydroelectricity remains a major activity of the TVA, there currently being 50 dams with a combined power output of 6000 MW.

Terrestrial energy

Energy radiated from the Earth and atmosphere into space.

Tesoro Corporation

Refiner and marketer of petroleum products in the western US handling over half a million **bbl** per day. The organisation was founded in 1968 and its earliest activities were in Alaska where, the following year, it opened that State's first refinery. There were acquisitions over the following decades one of the most important of which was the Golden

Eagle refinery in California. Now based in San Antonio Tesoro continues to expand and has in early 2007 entered into an agreement whereby it will take over one of the Shell refineries in California.

Tetralin

Tetrahydronaphthalene, molecular formula $C_{10}H_{12}$, a hydrogen-donating solvent used in the conversion of **brown coal** to liquid fuel. Tetralin is also used to assess, on a laboratory scale, the suitability of a particular brown coal for such conversion. The coal in finely divided form is heated with tetralin in an autoclave and the extent of conversion determined.

TEU

See **twenty-foot equivalent units**

Texas, lignite in

Lignite is currently being mined at a level of about 50 Mt yr^{-1} in northeast Texas and almost all of it is used in power generation. Like the **brown coals** of the **Latrobe Valley** it is mined at open cuts. It is not always realised that Texas is rich in lignite. The only other two states of the US with major production are very distant from Texas: North Dakota and Montana. Notwithstanding this resource, Texas imports large quantities of higher rank coal, largely from Wyoming, being the largest consumer of coal of all the states of the US.

Work has begun on a power facility at Bremond, about 100 miles from Dallas, which will produce 1720 MW of electricity using locally won lignite as fuel.

Texas, wind farms in

Texas has 17 wind farms with a combined generating capacity of 1400 MW. At the present time, there is development of wind power off the Gulf Coast where winds are powerful. Electricity generated there will be largely destined for Houston. This latest undertaking, which will use larger wind turbines than most of those currently in service, has been dubbed the '**Spindletop** of the 21st century'.

Textile waste, fuel use of

As with plastic waste, there is some interest in co-combustion of textile waste with coal as **pulverised fuel**. This is currently taking place in Germany although the very finely balanced economics of coal/textile waste co-combustion appear to have precluded its expansion. Not all textile waste is suitable for burning. Wool is not as it is a fibre protein material. It enters a fluid phase after ignition, with the result that a large proportion of the initial weight remains as a char-like residue.

Thar desert

Location of a coalfield in Pakistan discovered in the 1980s, hugely raising the known reserves of coal in that country. The coal is low-rank, being **lignite** with a moisture capacity as won of about 45%, significantly lower than that of some other such coals in the bed-moist state. The total recoverable resource is 1.75×10^8 Mt, the overburden being about 200 m in depth. Surprisingly, commercial use of the Thar coalfield has not begun at the time of writing. There are planned schemes to use it for power generation by burning it as **pulverised fuel**. Manufacture of **briquettes** is also planned. There is mining of coal, usually only of local significance, in other parts of Pakistan. For quality coal for the iron and steel industries, the country relies on imports from Australia.

Thermal analysis

A genre of laboratory techniques whereby a sample of the material of interest, perhaps powdered coal, is subjected to a (usually linear) temperature programme and one or more properties of the material determined. The simplest is thermogravimetric analysis (TGA) where weight loss during the heating is determined. This provides an instrumentational means of carrying out **proximate analysis** of coals. The first derivative of the weight loss as a function of time is usually also determined, this being referred to as the DTG signal. Another thermal analysis technique is differential thermal analysis (DTA) where the temperature of the sample, relative to that of an inert material such as silica experiencing the same temperature programme, is measured and recorded. Such measurements are useful, for example, in comparing the oxidation reactivities of a range of coals or in ascertaining at approximately what temperature a particular coal ceases to release volatiles. There is also thermomechanical analysis (TMA) which is a form of dilatometry enabling expansion and contraction effects to be measured quantitatively across a temperature range. This has been applied to predictions of the behaviour of coal ash deposited on the interior surface of combustion plant. Thermal analytical techniques such as those outlined above are used not only in fuels and combustion technology but also in many other areas including polymer science.

Thermal coulomb

Unit of entropy equal to 1 J K^{-1}. The term has its origin in the widely known analogy between heat transfer and electrical current flow, as fully explained below.

$$\text{Heat: } q = \frac{\Delta T}{R_{\text{th}}}$$

Electricity: Power (Watts), $P = I\,V$

where q = rate of heat transfer (W), DT = temperature difference (K), R_{th} = thermal resistance (K W^{-1}), I = current (amp $\equiv$ C s^{-1}), V = potential difference (volt $\equiv$ J C^{-1})

Analysing in terms of the units,

Heat: rate of thermal energy transfer, is measured in K/(K W^{-1})

which is equivalent to J s^{-1} K^{-1} K $\equiv$ (J K^{-1}) K s^{-1}

$$\equiv \text{J s}^{-1} \equiv \text{W}$$

Electricity: power is measured in (J C^{-1}) C s^{-1} $\equiv$ J s^{-1} $\equiv$ W

On comparing the two sets of units above, a similarity can be seen. The analogue for the part of the electrical equation in parentheses is J K^{-1}, the dimensions of entropy and defined the 'thermal coulomb'.

Thermal delivery

The rate at which gas is supplied to a **burner**, expressed in terms of the energy which the gas will release when burnt. For a simple Bunsen burner using natural gas it is about 40 GJ m^{-2} hour^{-1} where the area is referred to the open end of the burner column.

Thermal efficiency

A measure of the efficiency of converting a fuel to energy and useful work.

Thermal NO$_x$

NO$_x$ (NO/NO$_2$) arising from reaction between the nitrogen and the oxygen in the combustion air. It occurs at temperatures in excess of about 1200°C. Two chemical mechanisms for its formation are known to take place: the **Zeldovich mechanism** and the **prompt NOx** mechanism.

Thermal removal of methane from natural gas hydrates, EROEI for

The only method as yet at all tested for removal of natural gas from **natural gas hydrates** is depressurisation. However, it is readily shown by simple heat balance that melting of the ice to release the methane under circumstances where no energy was required to bring the hydrate and heating fluid into contact would be around 15, which is about the same as that for a typical conventional natural gas onshore.

The hydrates do of course occur at great depths at inhospitable locations therefore vast energy costs are required in accessing them, but these can be offset over a period by the good EROEI once production of methane commences. There is the further advantage that because the melting is at a low temperature the **quality of heat** of the fluid need not be high.

Thermal waste treatment
See **incineration**

Thermoelectricity
The production of electricity by the direct action of heat or through the direct conversion of heat into electricity e.g. a thermocouple.

Thisted, Denmark
A location of **combined heat and power** with **municipal solid waste** as fuel. Steam raised by the fuel is used in a **Rankine cycle** with superheating and the water at the termination of the work-producing step of the cycle is at 80–90°C, therefore having sufficient **quality of heat** for subsequent use as hot water. Power generation is at 3.3 MW.

tHM
See **tonne of heavy metal**

Thorium
Chemical element, symbol Th. Though not fissionable itself, on encounter with thermal neutrons, Th^{232} forms U^{233} which is fissionable and indeed produces more neutrons per neutron absorbed than U^{235} does. When Thorium (as ThO_2) is used as a nuclear fuel, U^{235} is present to initiate the process by providing neutrons. There has as yet been use of thorium in the nuclear generation of power but only on a trial basis

at places including **Julich, Germany**, **Obrigheim, Germany** and Kamini, India. India in particular is looking to thorium as a nuclear fuel with which to expand its power generating capacity.

There are major thorium reserves in countries including Australia, India, Norway, the US and Canada.

Three Gorges Dam, China
The largest **hydroelectric** power generator in the world. Construction began in 1994, and the dam began filling in 2003. Capacity by projected completion in 2009 is expected to be > 18 000 MW.

Timbalier Bay, LA
The first offshore oil production in the world was off the coast of Louisiana in 1945. In September 1948, a hurricane occurred at Timbalier Bay which inflicted significant damage to the offshore oil platforms. It was the second hurricane in the Gulf since offshore platforms came into use: another in 1947 did not damage the platforms.

Tinicum Marsh, PA
Wildlife reserve at Tinicum, also referred to as the John Heinz National Refuge, and home to some threatened species. In 2000, 192 000 US gallons of oil leaked from a cracked pipe operated by the US company **Sunoco,** causing significant damage at the reserve. The company was fined $2.7 million and damages of $865 000 were awarded against it.

Titan Neptune
A **supertanker** built in 1988 and sometimes chartered by Exxon to take crude oil from the Persian Gulf to Asian markets.

Other such vessels used by Exxon are *Asian Jewel* (built 1992), *Lysaker* (built 1989) and the *Malibu* (built 1989). All of these vessels are getting long in the tooth and, being single-hulled, will not in any case be allowed to operate worldwide for very many more years because of requirements imposed by the US and the EU that oil-bearing tankers have double hulls. The use of a single-hulled vessel rather than a double-hulled one, while such use remains legal, saves hundreds of thousands of dollars on a voyage from the Persian Gulf to Asia.

The oil company Total uses only double-hulled tankers

and Chevron are moving increasingly to such tankers. Conoco always use double-hulled vessels in US waters, anticipating by 10 years the legal requirement to do so.

Tokai (or Tokai Mura), Japan

Scene of an accident at a nuclear fuels processing plant in 1999 in which two workers died and one other received radiation at a harmful level. The plant was processing uranium for use at the experimental fast breeder facility at Joyo.

Tokuyama Refinery, Japan

In the years following recovery from WW2, during which there was expansion in many Japanese industries including car manufacture, Japan coped with its lack of indigenous fuel by importing more and more crude oil for domestic refining, necessitating an expansion of very limited refining facilities. The first **grass roots refinery** was the Tokuyama Refinery, which began production in 1957 and was operated by Idemitsu, a company which had been trading imported oil in Japan since 1911. A second **refinery**, at Hokkaido, followed about six years later. Crudes for processing at the two refineries operated by Idemitsu were received initially from Iran and a little later from the then Soviet Union.

What is now the Idemitsu Group of companies has retained as its primary activity hydrocarbons and petrochemicals, having diversified into numerous related processes and operations and being represented in very many countries. It is the third largest oil refiner in Japan, the largest being Nippon Oil Corporation which, like Idemitsu, has a refinery at Hokkaido. A typhoon in 2004 halted crude oil deliveries to both Hokkaido refineries.

Tokyo Waterfront Wind Power Generator

Also known as Tokyo Kazaguruma, two 70 m high wind turbines are located on reclaimed ground in Tokyo Bay. The two Kazaguruma generate approximately 2.5×10^6 kWh of electricity annually. This is equivalent to the annual electricity consumption of 800 ordinary households and affords a CO_2 reduction of 1700 tons.

Ton Chan

The scene of offshore oil fields in Thai waters being developed by Total with 2009 as the target production commencement date. They are in the same block as Bongkot where major gas production is already taking place, there also being significant condensate. An abundance of gas for power generation in that region is a step towards an eventual common grid for Thailand, Burma, western Malaysia and Singapore.

Tonga, energy scene in

This country in the Pacific, which with Fiji and Samoa comprises the region known as Polynesia, is highly primitive in terms of its energy supply, there being a reliance on wood and on **coconut waste**. Electricity for the centres of population including the capital Nuku'alofa is generated thermally using imported petroleum fuels. Private electricity companies can receive licences to generate and distribute electricity in the more remote regions.

tonne of Heavy Metal (tHM)

Used as the weight basis to express the performance of a nuclear reactor and as part of the unit for **burnup**.

Total

A multinational energy company and the fourth largest publicly traded integrated oil and gas company. Total operates in more than 130 countries and has over 95 000 employees.

Total's oil and natural gas production is approximately 2.5 million barrels of oil equivalent per day, and reserves stand at more than 11 billion barrels of oil equivalent. It produces oil and gas in 29 countries. The company is also a world leader in downstream activities, being one of the largest refiners in Europe and Africa.

Total Raffinaderij Antwerpen

The largest refinery operated by Total (the world's fourth largest oil group) and the second largest refinery in Europe. Crude oil is obtained from Rotterdam and is carried to Antwerp by pipeline for refining. The Fina Antwerp Olefins plant is adjacent to the refinery and receives from it feedstocks including **naphtha**.

Town gas

A generic term for manufactured gas supplied to homes and industry.

Trace element

A chemical element present in very small quantities (of the order of 10 ppm or less) in a material. Trace element leaching from **fly ash** is of particular concern with elements such as mercury and arsenic, being able to escape into the environment under certain conditions.

Trap

The trap is a configuration of **reservoir** and cap rock which prevents oil or gas from migrating and escaping laterally or vertically. Earth movements lead to the development of subsurface environments which are ideal for **hydrocarbon** accumulation. Through deformation or lateral changes in the sedimentary basin, hydrocarbons become trapped in either structural environments (anticlines, domes, salt-induced structures, tilted fault blocks and buried hills) or stratigraphic environments (unconformities, pinch-outs and permeability traps).

Trat field

Non-associated gas field in the Gulf of Thailand, producing daily 1.9 million m^3 of gas and 2100 **bbl** of condensate. There is at present one platform in the field, operated by the Uncol Corporation, which receives gas from eighteen wells. The gas is for local use.

Trees, rate of removal of carbon dioxide by

The value for a single tree used in carbon dioxide 'accounting' in the northern hemisphere is one tonne of carbon dioxide in 40 years. Hence, to offset a release of one tonne per year, 40 trees must be planted. Some trees, notably in the Sequoia forest in the US, are known to have existed for about 2000 years. Such a tree will therefore have removed 50 tonnes of carbon dioxide from the atmosphere.

Tremembé, Brazil

The location of two major shale retorting facilities. The daily output of the larger, called the Irati Profile Plant, is 3870 **bbl** of shale oil per day, plus by-products including shale gas which is liquefied for transportation to purchasers.

Trinidad and Tobago, oil and gas reserves in

Of all the nations in the Caribbean, this is the richest in hydrocarbons, there being both on- and offshore oil production at a total rate of about 160 000 **bbl** per day, over 80% of it from offshore. BP, who are strongly represented, obtain only offshore oil. The Trinidadian company Petrotrin obtain about 60% of their oil from offshore. Other firms including BHP are active in the region. Some **associated natural gas** is re-injected to the well, the remainder being collected for sale.

There are also non-associated gas fields and exploration of these is underway by organisations including British Gas. The largest such field is the Dolphin field with estimated reserves of 27 **bcm** of dry gas. A few miles from the Dolphin field is the Starfish 1-X field, jointly discovered by British Gas and Texaco only a few years ago. A condensate field has also been located which is capable of producing just under a million m^3 per day of gas with 430 bbl of condensate accompanying it. These figures represent 6–7% by weight of condensate, a higher than usual amount. Though it receives some Canadian crude oil, Trinidad and Tobago is a net exporter of both oil and gas. This includes some local export of gas as **liquefied natural gas** to other Caribbean countries. Obviously export of gas will increase when development of the newly discovered fields gives way to production.

The country also has a significant deposit of **bitumen** at **La Brea**.

Troll field
The largest gas field in Europe lies in the Norwegian sector of the North Sea, having over 100 wells. The field also contains large amounts of oil and is the scene of sub-sea separation of water from hydrocarbons, currently a rare procedure. The structure of the field is unusual in that there is a layer of oil, an aquifer and gas beneath that. Water inevitably accompanies in high proportion any oil/gas released from one of the wells at the field. After well exit, the water is separated from the oil/gas by means of a centrifugal pump provided with power from an **umbilical cable** and is re-injected into the well. The oil and gas go to a floating production unit (FPU) for further separation. The current level of oil production with sub-seawater separation is 60 000 **bbl** of oil per day plus **associated gas** with 40 000 bbl of water being re-injected.

Tuban, East Java
The site of the tenth oil **refinery** to be built in Indonesia. **Sinopec** has a significant holding. The capacity when the refinery has been commissioned will be 200 000 **bbl** per day, the crude being pumped from fields currently occupied by Exxon and Santos. Later crude from the currently untapped **Cepu field** will go to the refinery at Tuban.

Tug Hill, NY
Location of the Maple Ridge **wind farm** which has just entered operation with a capacity of 300 MW. The State of New York is aiming at 25% of power from renewables by 2010. Maple Ridge is the largest such facility east of the Mississippi.

Tullow Oil

Company with exploration activity in the southern North Sea and also in a number of African countries as well as in India, Bangladesh and Pakistan. From being purely an exploration company, it expanded into production on acquisition of certain BP gas wells in the North Sea, together with a share of the infrastructure including the terminal at Bacton in north Norfolk. It has also taken the newly discovered Munro field in the North Sea all the way from drilling to commercial production using reserve infrastructure capacity at adjacent fields.

Turbine

Steam turbines have been continuously developed in the UK since the early 1880s. They are the most important prime movers worldwide for electricity generation. Unit sizes vary from low power output to > 1300 MW_e. Steam is supplied at high pressure and temperature to the steam turbine. The energy in the steam is converted into mechanical energy by expansion through a series of fixed blades or nozzles and moving blades. A row of fixed blades together with its associated moving blades is known as a turbine stage. The fixed blades are attached to the turbine casing which contains the steam pressure, and the moving blades are attached to the turbine rotor which turns the generator to produce electricity.

Turkey, oil reserves in

These are fairly sparse and production is now 49 000 **bbl** per day, having been almost twice that fifteen years ago. Domestic oil therefore only accounts for 10% of the consumption, there being imports from countries including Saudi Arabia, Iraq and the Former Soviet Union. Turkey also has very little domestic natural gas. Exploration for offshore oil and gas in the Black Sea and the Aegean Sea is planned. The **Baku-Tbilisi-Ceyhan Pipeline** terminates in Turkey on the Mediterranean coast.

Turkmenistan, gas reserves of

Having a border to the south with Afghanistan and to the west with Iran, this country (which was once part of the USSR and is now part of the Commonwealth of Independent States) has massive non-associated gas reserves, about 2.7 trillion m^3 with prospects of more on further exploration. In the period January–August 2005, gas exports from Turkmenistan were 30 **bcm**. There are also oil fields with production at the level of around 100 000 **bbl** per day.

Tuxpán

The location of the first oil production in Mexico, in 1901. Export of oil from Tuxpán began in 1911. One of the participating companies was Pearson, now a huge publishing and media group. Pearson had involvement in Mexican oil from 1901 and traded as Mexican Eagle. Shell and Standard Oil were among the other oil companies active in Mexico at that time.

Twenty-foot equivalent unit (TEU)

A measure used to match vessel (including oil tanker) dimensions with berth capacity at ports and terminals. Although non-SI, the unit is widely used by organisations such as Lloyd's Register. The oil terminal at Galveston, Texas, can handle up to 125 000 TEU per year, an alternative measure to **bbl** or tonnes of oil per year.

Twistdraai

Coalmine near Johannesburg. Its most interesting feature is the **coal cleaning** that takes place there. Cleaning is in preparation for export and the **middlings** are used by **Sasol** for the preparation of **synthesis gas**.

Tyre derived fuel (TDF)

In the US, TDF is used as the primary fuel at facilities such as cement manufacture and pulp/paper manufacture. The number of tyres currently disposed of in this way in the US is equivalent to the human population. There is also use in power generation on an intermittent basis as at **Bay Front, Wisconsin**.

U

Uch Power Station
Gas-fired facility situated in southwest Pakistan, with a capacity of 586 MW and owned jointly by US and UK investors. In recent years its operation has been affected by the destruction of pipelines to the facility by anarchists.

Uganda, energy scene in
The country's major energy assets are its hydroelectric installations at Owen Falls, Kabale and Kasese. Currently, however, the total generation of power for the country is only about 500 MW. There has been erratic pricing of electricity, including a 536% increase in 1987. Uganda imports petroleum products. In the 1980s several major companies including Petrofina, Shell, Exxon and Total were exploring for oil in Uganda with World Bank financial aid but withdrawal due to eventual loss of interest result of which Uganda continued to be dependent upon imported oil. There is the further disadvantage that Uganda is a landlocked country therefore oil imported from outside the African continent has to be brought to Uganda by road or train usually via Tanzania. Hopes for a pipeline through Kenya to Uganda remain unfulfilled because of uncertain relations between the two countries.

Consequently wood fuel and charcoal together supply over three-quarters of the energy needs of Uganda, not only for homes but also for industries such as brick manufacture.

Ujung Pangkah field

Offshore gas field in Indonesian waters having recently been developed by a number of companies including Dana Petroleum, Amerada Hess and ConocoPhillips. Gas production at the field began in late 2006; the eventual rate of production is expected to be just under 3 million m^3 per day. All of the gas there is destined for electricity generation in East Java.

UK coal industry

Once a key player, producing 300 Mt yr^{-1} of **hard coal** 80 years ago, which was equivalent to three-quarters of the US production at that time. Serious decline began in the 1960s and by 1999 the output was a mere 37 Mt yr^{-1}. The industry was privatised about ten years ago. There is now limited activity in deep mining and in open-cast mining. Some coals from currently worked UK fields are too high in sulphur to be competitive in current international markets. A renaissance of the UK coal industry is not impossible if there is a swing to coal, coal-derived liquid fuels or manufactured gas, as vast proven reserves of coal remain in the UK.

UK electricity production

In the UK, electricity-generating capacity exceeds demand. Generation in 2002 was at 41 GW, almost 80% from thermal sources using conventional fuels. Nuclear power stations account for about 18%. Since 2002 there has been a marked increase in power generated at UK **wind farms**, and recently the total rate of electricity production at wind farms in the UK crossed 1 GW ($\equiv$ 0.15 **Million Barrels per Day of Oil Equivalent, MBDOE**).

UK net oil export

The UK is a net exporter of oil due to the quantity of reserves in the North Sea. In 2004, 0.22 million **bbl** per day were exported.

Ukraine, energy scene in

This country of population 48 million has 395 million **bbl** of proven oil reserves but is not at the present time self-sufficient, importing 350 000 bbl per day. This is in spite of a dramatic—almost 50%—decline in oil consumption since the early 90s. The country also has large natural gas reserves, about 10^{12} m^3, although actual supply has fallen short of demand and the Ukraine imports about three-quarters of its natural gas. There are 38 **short ton** of **hard coal** in the Ukraine, spanning the entire scale of rank from **anthracite** to **lignite**. Ukraine is also a net importer in this area, receiving from other countries about 6% of its hard coal. Electricity generation is at 50–55 GW with oil, natural gas and coal accounting for 60% and nuclear fuels the remainder. The Ukraine has an excess of electricity and exports at TW level. The country's transmission system is in bad need of renewal.

There has been, for about the last five years, significant **coal bed methane** recovery in the Ukraine, at the Volyn coal field in the west of the country. EuroGas is the operator.

Ultimate analysis

Analysis of a coal or other solid fuel according to the elements: C, H, N, S, O. Though there are traditional bench chemistry methods for such analysis, instrumentational approaches are now much more common. Whatever the method, oxygen is always determined by difference. Attempts to develop a method of direct determination of oxygen have involved exposure of the coal under inert conditions and a high temperature to a highly reactive carbon substance. The expectation is that this will remove the coal oxygen, which can then be determined as oxides of carbon. This approach has not, in exploratory work, met with sufficient success to justify its adoption in routine work.

Ultra-deep gas

Offshore gas with a well depth of 800 m or more: generally, such gas is in shallow water and so is not classed as deep-sea gas. It is the well with its surface at the seabed that is deep. At present there is major exploration for ultra-deep gas off the Louisiana, Mississippi and Alabama coasts. Exploration sites for contract have attracted many more bids than there are sites: 651 and 428 respectively in March 2005, from 72 companies.

Ultra-superheated steam

When a hydrocarbon fuel is burnt in a synthetic air mixture of oxygen (21%) and steam (79%) using a conventional burner, the high temperature flue gas leaving the burner is a mixture of steam and carbon dioxide known as ultra-superheated steam (USS). With the burner operating at a stoichiometric combustion ratio, the USS is oxygen free:

$$O_{2(g)} + H_2O_{(g)} + Fuel \rightarrow H_2O_{(g)} + CO_{2(g)}$$

The USS steam flame can be considered a 'designer flame' whose temperature and composition can be varied to suit specific process needs, such as **gasification** of waste plastics for example.

'Umbilical cable'

Figurative term for a sheathed cable which conveys electrical power, generated onshore, to where it is required in hydrocarbon operations below the surface of the sea, for example admittance of liquid hydrocarbon to a tanker at a position where its hull is immersed in the seawater. Such power is usually transmitted at up to 1000 V. The longest such cable at the present time is in Norwegian waters and extends to 210 km.

Underbalanced drilling

In conventional (overbalanced) drilling for offshore oil, the drilling fluid provides a positive pressure with respect to that in the hole so that hydrocarbons do not leak out during the drilling process. It sometimes happens that the pressure of hydrocarbons at the scene of drilling is so high that it cannot be contained by the drilling fluid. In such circumstances, instead of containing the hydrocarbon in the hole, the fluid regulates its exit rate and prevents blow out i.e. unrestricted exit of hydrocarbon. A triumph of developments in underbalanced drilling was the recent introduction of oil production at the **Clair field**.

Underground gasification of goal (UGC)

The feasibility of making gas by gasifying coal *in situ* was investigated independently in the UK and Russia in the 1920s–30s. UGC is air-blown, not oxygen-blown, therefore we might at first expect only to obtain a form of **producer gas**. However one of the advantages of UGC is that the carbon monoxide obtained by the chemical reaction of the coal with oxygen is accompanied by a significant amount of **coal bed methane** and this raises the calorific value to ≈ 11 MJ m^{-3}. A gas of this **calorific value** would contain about 35% CO, 20% CH$_4$ and the balance nitrogen. The deeper the site of **gasification**, the greater the methane content.

Currently, there is only exploratory activity in the UK, in places such as the Lothians region of Scotland. It is envisaged that gas produced by UGC could be burnt on site to make electricity and the carbon dioxide captured and put back into the mine. There are many 'idle' coalmines in the UK and UGC is seen as a way of utilising this resource in an environmentally sound way. In the US, promising new developments in UGC (including the use of oxygen instead of air, to eliminate the diluent nitrogen in the product gas) were made in the 70s and 80s but they did not lead to new facilities because natural gas was cheap at that time.

United Arab Emirates (UAE), hydrocarbon reserves and production

Like the other oil-producing countries in that region of the world, the UAE was a late entry into the hydrocarbon scene. The first export of oil from Abu Dhabi was in 1962. Of the four other emirates which with Abu Dhabi constitute the UAE, only Dubai is a major hydrocarbons producer, yet its output is a twentieth or less of that of Abu Dhabi. The estimated reserves of oil for the UAE are 98 billion **bbl** and discoveries continue. There is both onshore and offshore production. Output is about 2 million **bbl** per day, most of which is exported, largely to Japan. Japan itself, in the form of the company INPEX, has an investment in the Upper Zakum field in waters off Abu Dhabi, which is the largest oil field in the UAE and produces about half its total output.

The UAE has also been producing natural gas since 1967, proven recoverable reserves being 6000 **bcm**. Again, Abu Dhabi leads with Dubai a distant second. There is **associated natural gas** and also condensate fields. There are **liquefied natural gas (LNG)** plants in the UAE for export of the gas. One of the newer operations in the UAE hydrocarbons industry involves taking gas from a non-associated field, removing the **natural gas liquids (NGL)** and condensate and re-injecting the gas itself into the well. The target production of this enterprise is 125 000 **bbl** per day of condensate and 12 000 Mt per day of NGL.

There are facilities in the UAE for making LNG, largely destined for eventual use in the Far East.

United States Air Force (USAF), use of renewables by

The USAF has become an exemplar for American enterprises in its use of renewables to provide electricity for several of its major air bases. USAF purchases wind power but also generates some itself, at Ascension Island in the Atlantic and at an air force base in Wyoming. There is also significant activity in power from biomass combustion and in **pumped storage of electricity**.

Unleaded petrol

Petrol (gasoline) that does not contain **tetraethyl lead** as an antiknock agent. **Methyl tertiary-butyl ether (MTBE)** and other non-lead containing compounds are used as additives to produce the same effect.

Unsaturated hydrocarbon

Organic compounds containing carbon and hydrogen where the carbons are attached to each other via double or triple bonds; see **olefins**.

Uranium, availability of as a nuclear fuel

The country with the largest proven reserves of Uranium (in the form of U_3O_8) is Australia. There has been export of uranium from Australia for 25 years with the proviso that it is used for electricity generation only (minute amounts are used in the preparation of medical isotopes), and that precise records are kept of amounts received and used by the purchasing countries. Nations currently receiving uranium from Australia include the US, Japan and several members of the EU including Sweden and Great Britain. Other countries producing uranium, sometimes just for domestic use, include France, Germany and Spain, each of which also imports uranium. The former Soviet Union is a major producer, as are Brazil, South Africa and Namibia. Canada produces over four times as much uranium as the USA.

Uranium 233 (U^{233})

Fissionable isotope of uranium. The precursor to U^{233} in a nuclear reactor is **thorium**.

Uranium 235 (U^{235}), fission of

When a neutron is captured at the nucleus of this isotope of uranium, the energy is divided among the protons and neutrons and it becomes destabilised. The nucleus consequently undergoes fission to form lighter elements. There are many such possible nuclear reactions, and the mass number of lighter elements yielded is in the approximate range 75–160, the distribution being bimodal. Many of the nuclear processes release neutrons, commonly two or three, and these enter other U^{235} atoms and a chain reaction takes place. One such process is:

$$U^{235} + n \rightarrow Ba^{141} + Kr^{92} + 3n$$

where n denotes a neutron and Ba and Kr are the chemical symbols for Barium and Krypton respectively. The heat released by the above process is 197 MeV (mega electron volts), equivalent to 3.2×10^{-11} J. The combustion of methane at the level of one molecule is considered below for comparison.

CH_4 + air $\rightarrow CO_2$/water

889 kJ per mole of methane

889 000/N_o J per molecule of methane, where N_o is Avogadro's number, 6×10^{23}

$\Downarrow$

energy released per mol methane reacted = 1.5×10^{-18} J

One U^{235} atom undergoing fission releases 20 million times as much heat as a molecule of methane reacting chemically. U^{235} comprises only 0.7% of naturally occurring uranium and this has to be raised to 3 to 5% by enrichment in the preparation of the nuclear fuel. Uranium so enriched is used in the form of pellets of uranium oxide about an inch long, and these are incorporated into **fuel rods**.

In addition to the type of process outlined above, transuranic elements heavier that U^{235} are formed during neutron capture. The most important such transuranic element is plutonium (symbol Pu) formed from neutron capture by U^{235} followed by β decay. These have the capability to undergo decay processes, and such processes are very powerful. Only quite a small proportion of the uranium goes along the path to plutonium, therefore plutonium decay accounts for a significantly disproportionate amount of the total heat release from the nuclear fuel. Consequently the 1% of plutonium that spent uranium fuel contains is recovered and used in the preparation of **mixed oxide fuel (MOX)**.

Plutonium production can be enhanced by neutron capture by the other, more abundant isotope of uranium U^{238} and this occurs in a **fast reactor** which uses fast neutrons. By contrast, the neutrons in conventional U^{235} fission are thermal neutrons, having equilibrated thermally with the atoms upon which they impinge. A reactor operating thus is a thermal reactor.

USA, recent drilling for oil and gas in

New oil and gas wells, on- and offshore, in the USA for 2006 totalled 49375. In 60% of the new wells, non-associated gas was being sought. This was the largest number of new wells for a single year since 1986. It is highly probable that President Bush's State of the Union address in January 2006, in which America's energy situation was addressed with some earnestness, was one reason for the intensified drilling endeavours.

US natural gas hydrate reserves

Believed to be sufficient to yield on dissociation a quantity of natural gas equivalent to 200 times the conventional natural gas reserves of that country. More than a tenth of this reserve is in the Gulf of Mexico and approaching a third of it is in Alaska or in waters off Alaska. There are other significant reserves off the Pacific coast, northern and southern, and along the length of the Atlantic coast. The belief that there will always be such hydrates at a continental shelf is exemplified by the US.

US total crude oil refining capacity

At the time of writing, the total crude oil refining capacity of the US is about 17 million **bbl** per day.

US total energy consumption

The total energy consumption of the US was estimated to be 100.3 quadrillion (10^{15}) **British Thermal Units (BTU)**, which is equivalent to 10^{20} J, in 2004. That was a 2% rise over the previous year. **Renewables** accounted for 6% of this figure.

US pre-1776 use of coal

Long before Europeans explored North America there was use of coal by the indigenous people of the southwest. A map from the 1670s shows 'charbon de terra' in north Illinois. By 1760, over fifteen years before the Declaration of Independence, coal had been discovered in Pennsylvania, Ohio, Kentucky, Virginia and West Virginia, all major coal producing states almost 250 years later.

At the time of independence, most of the coal for the original 13 states was imported from England or from Nova Scotia, although some came from Virginia.

Utah, coal mining accidents in

This state has had more than a proportionate number of serious accidents in the mining industry over the decades. One of the earliest was in 1900 when 200 lives were lost, some of them through the inhalation of **afterdamp** which followed an explosion originating in blasting powder. About a quarter of a century later, 173 lives were lost at another coalmine in Utah. Within living memory, accidents in Utah coalmines in 1958, 1963 and 1984 claimed the lives of four, nine and twenty-seven persons respectively.

Utility

The name given to a public service provider, for example electricity, gas, telecommunications, waste disposal and water companies.

Uzbekistan, CIS

Location of an **underground gasification of coal (UGC)** facility first set up during World War II and still in use today. It is one of the very few operating facilities in the world at the present time. There is a revival of interest in such facilities, including in the UK, but no actual production as yet.

Vanadium

Chemical element, symbol V. It is frequently found in oxidised form in the solid residue remaining from the combustion of fuel oils.

Vankor field

Oil and gas field in Siberia currently being developed. The area is isolated and lacking infrastructure. Target oil production is 300 000 **bbl** per day which will be transported along a 775 km pipeline to a terminal in China.

Vattenfall

Swedish power company currently undertaking one of the most ambitious ever **wind turbine** projects. Vattenfall's Baltic Sea wind farm with a capacity of 640 MW is expected to commence electricity production in 2010.

Vector

A tanker vessel, which, while bearing over 8000 **bbl** of refined petroleum material, collided with the ferry Doña Paz (Lady Peace) in waters off the Philippines in December 1987. There was ignition and loss of 4375 lives, the worst peacetime maritime accident ever. Doña Paz, which under another name had been written off in 1979 after a fire on board but later salvaged and returned to use, was significantly overloaded with passengers at the time of the collision. Both Vector and Doña Paz were totally destroyed in the accident.

Velva, ND
A site of **biodiesel** production using a seed crushing plant. The operators have plans to establish a presence on **Jurong Island** by 2007. This will give Singapore its first biodiesel manufacturing plant, with an output of 1–2 million **bbl** per year. The co-existence at Jurong Island of conventional oil and gas facilities and biodiesel facilities is comparable to the wind farm at Rotterdam. Each sends out the message that although the major oil producers are supportive of renewables, even if and when CO_2 reduction targets set at the **Kyoto Protocol** are met there will still be massive amounts of conventional oil and gas in use.

Ventspils
Port city of Latvia, for almost 50 years the site of a terminal for oil export from the former Soviet Union. It is also the terminus of a pipeline used to convey crude oil from Russia to Latvia for that country's own use. Latvia produces no oil although there are significant reserves and foreign licence holders are active in development.

Venture II
An advantage of a **floating production, storage and offloading (FPSO)** over a fixed installation is that it can be disconnected from the sub-sea wells in anticipation of a cyclone. This happened in January 2006 when the FPSO Venture II was operating off northwest Australia as Cyclone Clare approached. Venture II had a production of about 60 000 **bbl** of crude oil per day and according to *The Australian*, January 10th 2006, disconnection meant that 'output was merely delayed, not lost'.

Vestas
Danish manufacturer of **wind turbines** and one of the leading suppliers in the world. In 2004, the company delivered 2270 such turbines worldwide, having a combined capacity of 2784 MW which in operation gave carbon dioxide savings of millions of tonnes. Vestas have a presence in North America as Vestas-American Wind Technology Inc., and are a major participant at **Wild Horse**.

Vietnam
See **White Tiger**

Visbreaking

A process used to reduce the **viscosity** of refinery residues by thermal **cracking** to enable the products to be combined with a smaller quantity of valuable blending oils than would otherwise be required for heavy furnace oil manufacture.

Viscosity

Fundamental property of a substance which quantifies its resistance to flow. The dynamic viscosity, usual symbol μ, has units of kg m^{-1} s^{-1}. The kinematic viscosity, usual symbol ν, is defined as:

$$\nu = \mu/\rho$$

where ρ is the density in kg m^{-3}, and has units of m^2 s^{-1}.

The viscosity of a liquid fuel is one of its most important properties for utilisation purposes. Older literature uses the poise and the stoke as units of dynamic and kinematic viscosity respectively, and these are defined by:

$$1 \text{ poise} = 1 \text{ g cm}^{-1} \text{ s}^{-1}$$

$$1 \text{ stoke} = 1 \text{ cm}^2 \text{ s}^{-1}$$

and as these units are based on mks (i.e. based on metres, kilograms and seconds) and not on Imperial units, their conversion to SI is straightforward. Taking liquid water as a benchmark, its dynamic viscosity at 15°C is 10^{-3} kg m^{-1} s^{-1}. Some light organic compounds have lower values than this, for example cyclohexane has a value at the same temperature of 3×10^{-4} kg m^{-1} s^{-1}. A gasoline will have a typical value of 6×10^{-4} kg m^{-1} s^{-1} at room temperature. Petroleum fractions of higher boiling range than gasoline have higher viscosities than water, a kerosene being about twice as viscous as water. Residual fuel oils can have dynamic viscosities at room temperature of the order of 10 kg m^{-1} s^{-1}, a factor of 10^4 higher than that of water.

Viscosities are often measured by the time taken for descent of a column of the subject liquid under precisely controlled conditions. This is the principle of the Saybolt viscometer and of the Redwood viscometer. Such so-called viscosities are therefore measured in seconds and conversion to true units of viscosity is according to an established chart. This contains parallel vertical scales of the viscosity in seconds and the specific gravity (water = 1) at 60°F, and an inclined scale of the true viscosity in centipoise. A line is drawn between the viscosity in seconds and the specific gravity; where that line intersects the inciined scale is the viscosity in centipoise. This is a classical technique pioneered in about 1950 with extensive application in the 21st century.

The viscosities of liquids decrease with temperature, gasolines by a factor of two over a temperature interval of about 40°C. When two organic liquids are blended, the viscosity of the blend needs not only to be known but to be planned in advance so that the blending gives the required viscosity. This is particularly important in the preparation of lubricating oils. Empirical charts apply once more. The viscosity range of one type of oil on a logarithmic scale forms one vertical axis and that of the other oil on the same scale forms the other. They are separated by a horizontal scale from 0–100% composition in one direction and from 100–0% in the other. The viscosities of the two oils to be blended can be joined by a line, and this forms a plot of viscosity of the blend against composition.

Dynamic viscosities appear to be more prevalent in the literature for petroleum products than kinematic viscosities, although conversion from one to the other is trivial. The value of the kinematic viscosity (in some older texts called the kinetic viscosity) is twofold. First, a number of dimensionless groups used in heat transfer, including the Rayleigh number, contain the kinematic viscosity. The Reynolds number is often formulated in terms of ν rather than μ. Secondly, the kinematic viscosity, having units $m^2\ s^{-1}$, is the analogue for momentum transfer of the diffusion coefficient D for mass transfer and of the thermal diffusivity α for heat transfer. Each of these has units $m^2\ s^{-1}$ so the quotient is dimensionless. The best known such quotient is the Prandtl number Pr, defined by:

$$Pr = \nu/\alpha$$

Viscosity is one of a number of physical properties which characterise a particular liquid fuel. Others include density and **flash point**.

Vitrinite
A **maceral** group derived from the cell wall material (woody tissue) of plants, which are chemically composed of the polymers, cellulose and lignin. Matrix vitrinite is almost always the most abundant maceral present in coals and makes up the groundmass in which the various **liptinite** and **inertinite** macerals are dispersed.

Vitrinite reflectance
Vitrinite reflectance is related to the aromaticity of **vitrinite** in the coal. It increases with increasing rank, hence its use as an index of coal rank. The reflectance of vitrinite is the percentage of directly incident light that is reflected from a plane polished vitrinite surface, usually

under oil immersion. Values may be obtained using polarised or non-polarised light. Under plane polarised light, the vitrinite reflectance varies over 180 degrees between a maximum and a minimum. Usually, the mean maximum reflectance is reported for 100 particles. Under non-polarised light, reflections from all directions on the vitrinite surface are integrated to give the mean random reflectance value. This value will always be lower than the mean maximum reflectance of vitrinite. Determination of vitrinite reflectance is time-consuming and expensive, requiring highly trained personnel. The mean maximum reflectance values can vary enormously for the same sample between operators. The mean random reflectance can now be automated. This method, which is recommended, provides more reliable results.

VLCC

Very Large Crude Carrier of which the Hellespont Alhambra is the largest. Double-hulled, it has a capacity of 513 684 m³. VLCCs cost about $US125 million and one such carrier in a day will earn up to about $US75 000.

Volatile organic compounds (VOCs)

The term applied to a large number of toxic organic compounds that easily vaporise forming gases. Formed during combustion of coal, natural gas, oil, petrol, waste materials and wood, also released by glues, paints and solvents. VOCs react with oxides of nitrogen to form ground level ozone or smog.

Vung Tau

A beach resort in Vietnam severely impacted when a Vietnamese oil tanker and a Liberian registered one collided close to it in 2001.

W

Wabash River, Indiana

The scene of **British Thermal Unit (BTU) conversion** on a large scale. Local **coal** is gasified and burnt to drive two **turbines**: a gas and a steam turbine which produce electricity at up to 192 MW and 105 MW respectively. Although the process is integrated, the gasifier and the turbines are separately owned, and gas rated at 900 billion BTU per month is passed from one to the other.

Walnut shells

Walnut shell is sometimes made into solid fuel as **briquettes**. Less well known is the fact that very finely ground walnut shell is one ingredient of a sealing agent used in the drilling of oil wells.

Waste heat recovery

The process of capturing heat which would otherwise be lost to the environment, to heat water for another process in the plant or for district heating, for example.

Waste oil, re-use of

Waste oil from industrial sources and from garages can be taken to a waste oil bank, a facility operated by a local authority. From there, it can be sent for processing or processors of waste oil will themselves collect it. It can be prepared for re-use in one of the following ways. It might be used as received as a fuel, or it might be de-watered and filtered before being used as a fuel.

There is significant use of such fuel in the UK power industry. It is in fact quite an attractive fuel not only because it is inexpensive but because it tends to contain less sulphur than hydrocarbon material in the same boiling range produced directly as residue from crude oil refining. On the other hand, the waste oil might be re-refined in which case a base oil for lubricants is generated. To this base oil, other hydrocarbon liquid is added as necessary to adjust the **viscosity** and other properties upon which the performance of a lubricant depends.

There has been very limited activity in cracking waste oil with good results in terms of liquid products obtained which are suitable for refining into useful end products. The cracking of waste oil is however economically unfavourable when compared with the approaches described in the previous paragraph and is not likely ever to account for a significant proportion of the waste oil available.

In the UK at present time, about half of the waste oil generated is processed for future use as fuel or lubricant, while about 6% goes to landfills and the remainder is not accounted for, there being no statutory requirement to record the eventual destination of waste oil.

Waste to energy (WTE) plant

A combustion plant for the burning of **municipal solid waste (MSW)** to make steam and hence electricity. In the US, 15–20% of the total MSW generated finds its way to a WTE plant. The state with the greatest WTE activity is Pennsylvania.

A WTE plant in southern California, called the Commerce Refuse-to-Energy Facility, burns 360 tonnes of MSW a day, generating 10 MW of electricity for sale.

Water, contamination of by oil

A concentration of only 1 ppm of oil in water is sufficient to make it unfit to drink. That is why in disposal of oil by landfill methods, care has to be taken to ensure that it does not find its way into water supplies. This requires knowledge of the geological structure at the site of the landfill.

Watson field

Gas field in the North Sea believed to contain 15 million **bbl**. The field is distant from any existing infrastructure and so has not yet been developed, but two other stranded oil fields in the same quadrant of the North Sea (number 22) are also being investigated and it is hoped that eventual production will result.

Wave energy
see **Marine wave energy**

Waverly, TN
The location of an accident in 1978 involving **liquefied petroleum gas (LPG)** in which 15 persons died and 56 were injured. On the Louisville-Nashville railway line two LPG-bearing cars from a derailed train released their contents which exploded. Damage to the built environment in Waverly was extensive.

Weapons-grade plutonium, possible use of as a nuclear fuel
The residual plutonium in spent uranium nuclear fuel can be recovered and made into **mixed oxide fuel (MOX)** for further use in electricity generation. There have been proposals, both in the US and in the former Soviet Union, that in the decommissioning of nuclear weapons containing plutonium the element could be put to such use, becoming an ingredient of MOX for power generation. Some such plutonium has already been sent from the US to France for conversion to MOX and will be returned to the US for use. Such processing is also taking place in the US as **Savannah SC** and MOX produced there will from 2007 be used at two local nuclear power stations.

Well-to-wheels
Well-to-wheels analysis is a systems approach to assessing the energy consumption and **greenhouse gas** emissions associated with different fuels and vehicle propulsions systems. A well-to-wheels analysis takes into account energy use and emissions at every stage of the process, from the moment the fuel is produced at the well to the moment the wheels are moved. Using this type of analysis, a vehicle with a diesel powered internal combustion engine can be directly compared to a **fuel cell** vehicle that uses hydrogen made from natural gas, for example, both in terms of emissions and energy use. This is particularly important when considering hydrogen fuel cell vehicles since there are numerous ways to produce hydrogen, some of which are clean and efficient and others which are polluting and energy intensive.

West Sole field
Non-associated gas field in the North Sea, discovered by BP in 1965

and entering production in 1967, predating by about three years oil production from the **Argyll field**.

Western refining

Oil refiner and marketer based in El Paso TX, having a refining capacity of 117000 **bbl** per day and also considerable storage capacity. Distribution is both by pipeline and in containers.

Westphal balance

Classical, although not obsolete, means of determining the density of a petroleum fraction. Its use involves immersing a plummet first of all in water and adjusting the balance so that alignment of the beam supporting the plummet is obtained. The plummet is then immersed in the subject liquid, and small weights (riders) placed on the beam so as to restore the alignment it had when the plummet was in water. In effect, the device has measured the buoyancy force on the plummet due to the petroleum liquid relative to that on the same plummet due to water, so the density of the liquid relative to water is calculable, in fact, to four decimal places if riders of a wide range of size are used. The density obtained in this way will usually be converted to the **API gravity** scale.

Weyburn, Saskatchewan

Site of an innovative activity in which **enhanced oil recovery** and **carbon sequestration** are achieved in one operation. Carbon dioxide, obtained as an unwanted by-product of a **lignite** gasification process taking place across the US border in **North Dakota**, is taken to Weyburn along a 200 mile pipeline and injected into oil wells there in order to raise their yields of crude. Carbon dioxide, which would otherwise have been discharged into the atmosphere in North Dakota, therefore ends up stored in oil wells in Saskatchewan. Current figures are approximately 2.5 million m^3 per day (referred to 1 bar, 288 K) of carbon dioxide injected into the wells, resulting in an extra 5000 **bbl** of crude oil produced.

Whale oil barrel

Whale oil was widely shipped as illuminating oil before petroleum fuels became available. It was contained in barrels during shipment, as was crude oil until the late 1870s, when the first oil tanker (the **Stat**) entered service. This combination of circumstances has sometimes led to the view that the **barrel** in its present day sense, meaning 159 litres, was derived from the whale oil barrel. It is now known that this is not true. The whale oil barrel was about 20% smaller than the barrel, and was not in any case precisely standardised, there being variations of 5–10%. The whale oil barrel was often made by the ship's carpenter while at sea.

White oils

Lubricants for machinery used in the food and pharmaceutical industries, where any contamination of the final product can lead to human ingestion of the contaminant. Major oil producers including ExxonMobil, Shell and BP manufacture petroleum-based white oils. White oils have to conform to prescribed specifications. Criteria for classification of a particular white oil as 'food grade' include colour, odour and taste. Less subjective requirements are low sulphur content and low susceptibility to the formation of carbon particles by pyrolysis.

White Deer, TX

Location of a **wind farm** of 80 MW capacity. Operated by Shell, it occupies land also containing some oil wells and traversed by gas pipelines. The turbines were manufactured by Mitsubishi.

White gasoline

(1) Term applied to unleaded gasoline in the 1950s and earlier, when most gasoline was leaded.

(2) Material in or close to the boiling range for gasoline but not for automotive use. Applications include small combustion appliances such as camping stoves. White gasoline in this sense is similar to **Coleman fuel**.

Whitegate Refinery

Situated in Cork, Ireland and operated by ConocoPhillips, this facility produces 1000 **bbl** per day of **soybean** oil for blending with conventional diesel in order to reduce fossil fuel carbon dioxide emissions and help Ireland achieve its Kyoto targets. The blending of vegetable oils with mineral fuels in the mitigation of carbon dioxide build-up is of course entirely analogous to co-firing of biomass with coal. **Corn alcohol** can also be used in this way.

White Tiger

White Tiger, referred to locally as 'Bach Ho', is the name of a currently very productive oil field in Vietnam operated by Vietsovpetro. There is also the Black Lion ('Su Tu Den') field estimated to contain 250 million **bbl** of crude oil and now being developed. Vietnam is becoming a major producer of crude oil both for domestic use and export.

Whitedamp
Term used in mining as a synonym for **carbon monoxide**.

Wild Horse
Wind farm in Washington State, completion expected around December 2006. It will comprise 127 **Vestas** turbines each of 1.8 MW giving a total capacity of 229 MW. Also in Washington State is the 150 MW Hopkins Ridge **wind farm**. Of the representative examples of wind farms in the US given in this volume, several are in states having a border with Canada, where there is some concentration of activity in this means of power generation.

Wildcat
An exploration well for which a prior geological survey is either absent or incomplete. The depth at which oil will occur, if at all, is not known and this calls for extra measures including a denser **drilling fluid**.

Wilmington field
Oil field in southern California, discovered in 1932. Part of it was once operated by a consortium comprising Texaco, Humble, Unocal, Mobil and Shell, and the acronym THUMS is used by Occidental who took over from the consortium in 2000 at a price of $US 1.20 per proven **barrel** of oil. Occidental recouped its investment in 14 months. Operations are from four artificial islands in Long Beach Harbour, disguised and camouflaged to look like resort islands.

Wind farms, ecological effects of
In the US each year tens of thousands of birds are killed by **wind turbines**, the wind farm at **Altamont Pass, CA** having experienced this difficulty particularly severely. In evaluating this figure, however, it must be remembered that about a billion birds die accidentally in the US each year reasons such as building and vehicle impact. It is because threats to bird life from wind farms is small when compared to loss due to other factors that experts in countries such as Australia deprecate alarm, reminding the public that wind farms are established to protect the environment.

In the US, several species of bat have also been killed by impact with wind turbines.

There are also demands from some ecologists that effects of off-shore wind farms on marine life be evaluated. Some fish and marine mammals use the Earth's magnetic field as a means of orientation and electromagnetic fields from cables could interfere with this.

Wind power, total world power production from
The total world generation of power from wind is approximately 50 GW—more than the total electricity consumption of the UK at present. This is currently increasing by about 35% per year. If this rate of increase is sustained, 12% of the electricity produced in 2020 will be by wind power.

Wind power, capital investment required
In the UK, the amount of capital investment required for new plant per kW of capacity was about £750 at the end of 2005. (This is equivalent to £750 000 per MW.) There is, however, significant trade in second-hand **wind turbines** between mainland Europe, where there is a trend towards replacing existing turbines with larger ones, and the UK. The Nissan plant in Tyne and Wear, UK, will soon be meeting about 10% of its electricity requirements with wind power, via seven second-hand turbines. A further dimension to the economics of wind power is the cost of the copper conductors required for transfer to the grid. In Western Australia recently, a wind farm project was put on hold partly because of a surge in the price of copper.

Wind turbine
A device by which electricity can be generated from the wind. As with **hydroelectricity**, thermodynamic analysis is very simple as illustrated below.

A wind generator with a 25 m diameter blade has a cut-in (entry) wind speed of 10 m s^{-1} at which velocity the turbine generates 110 kW of electrical power. Determine the overall efficiency of the turbine-generator combination. Use a value of 1.22 kg m^{-3} for the density of air.

We take the exit speed of the fluid to be negligible (as we did when considering hydroelectricity) and consider the process to be adiabatic i.e.

$$W = \frac{m'c_1^2}{2}$$

where W is the rate of work done at the wind turbine (W), m' is the mass flow rate (kg s^{-1}) and c_1 is the cut-in (entry) wind speed (m s^{-1}).

In applying the continuity condition to determine m', the applicable area is that of the circle swept out by the fan blade.

$$m' = 1.22 \times 10 \times \pi \left(\frac{25}{2}\right)^2 = 5989 \text{kgs}^{-1}$$

Therefore,

$$W = \frac{5989 \times 10^2}{2} = 300 \text{kW and Efficiency} = \frac{110}{300} = 0.37(37\%)$$

Depending on their size and weight, rotation of the turbine blades is typically up to about 100 rpm. There are over 130 operational wind farms in the UK with more under construction and in the planning. The first was that at **Delabole**. A further example is Windy Standard Farm in southwest Scotland, which became operational in 1996. It consists of 36 wind turbines, each of 600 kW maximum output or a combined maximum power of 21.6 MW. There are numerous small facilities consisting of a privately owned single-turbine of 5 kW. It is used to power its owner's property, any excess being sold to the grid. One of the largest is planned for the island of Lewis; the initial proposal was for 234 turbines, but a revised application for 181 turbines is awaiting the decision of the Scottish Executive having gained local planning approval.

The capacity of the largest currently commercially available wind turbines is in the 2–3 MW range: such are being installed at **Bear Creek, Pa.** Those proposed for the **Cape Cod wind project** can produce at up to 3.6 MW. The minimum wind speed required for electricity generation is about 7 m s^{-1} (or 16 mph), a figure currently being used as a benchmark in constructing the facility at **Sir Baniyas Island** where wind speeds often reach 25 mph.

Possible hazards to low-flying aircraft due to wind turbines have entered the debate into the acceptability of a proposed wind farm in Scotland. A wind turbine taller than 300 ft (91 m) in the UK is required to have illumination so that aircraft pilots can see it. Military aircraft doing exercises sometimes fly lower than that.

Winkler process

A means of making **blue water gas** or **producer gas** in a **fluidised bed**. A common starting material is **char** made from **brown coal**.

Wobbe index

For a fuel gas,

$$\text{Wobbe index} = \frac{\text{calorific value (MJ m}^{-3})}{\sqrt{\text{density (air = 1)}}}$$

Hence for natural gas, the Wobbe index is about 50 MJ m^{-3}. The Wobbe index is used when considering the interchangeability of a **burner** from one gas to another.

Wood, fuel use of

Wood has been used as a fuel throughout history and it was only in the late 19th century that coal overtook wood as the primary solid fuel in the US. Since then, the unit of fuel wood for sale in the US has been the cord, a cube of 128 ft^3. There has been much revival of fuel use of wood in recent years because of its **carbon neutrality**. In domestic fireplaces there is a strong aesthetic appeal of a wood fire.

One has to distinguish between wood grown with fuel use as its intended purpose, sometimes called fuel wood, and wood waste, for example from sawmills, diverted to fuel use. Categories of fuel wood include log wood and wood chips, whereas **pelletised wood** and **briquettes** are made from waste wood. Finland and Sweden are world leaders in fuel utilisation of wood. Seasoning—air drying over a period of weeks or months—is necessary with such products to bring the moisture content down to about 12–15%. It is sometimes the practice to leave the leaves on the wood during seasoning and remove them later, as their surface area assists in moisture loss and accelerates seasoning. Seasoned wood has a **calorific value** of approximately 17 MJ kg^{-1}. Wood combustion characteristics differ in some degree between species. Oak and elm, having a higher density than many woods, are sometimes better mixed with a more easily ignitable wood such as beech. Spruce and horse chestnut logs can split open suddenly during burning, creating a hazard if the hot debris is able to exit the fire. Trees of the genus *Eucalyptus* have found only limited application as fuel wood. They originate in Australia but have been naturalised in places such as California.

Wood is the starting material in **coalification**: reactivity decreases with rank and accordingly wood is more reactive than peat or any of the successive ranks of coal. Wood starts to undergo **devolatilisation** at temperatures below 100°C, an indication of its high reactivity. When a coal has a high ash content, it is collected during the coalification process: the ash content in the initial vegetation deposit is low. Consequently wood fuels themselves are low in ash and this is a significant advantage.

Africa and Asia jointly account for ≈70% of the wood grown for fuel use; there is no production at all in the Middle East. The Americas and Europe account for ≈20%. If fuel wood and wood waste are considered together, sub-Saharan Africa has the highest *per capita* usage, most of it in applications considered primitive in the extreme.

Wood gas

The analogue for wood of **retort coal gas** for coal, made by heating wood to produce a fuel gas with charcoal and tar/liquor as side-products. The **calorific value** of wood gas is about 12 MJ m^{-3}, that is, about 40% lower than that of retort coal gas.

In Ireland at the present time there is **combined heat and power** from wood gas, using **gensets** actually operated by spark ignition (not compression ignition) engines. The effluent gases from the gas combustion are heat exchanged, some of the heat being fed back to dry the wood starting material in readiness for gasification. There is however some heat remaining for the provision of hot water.

Wood River, IL

The location of a ConocoPhillips refinery, the largest operated by that company. Its capacity is 32 000 **bbl** per day of crude oil and is currently installing a gasoline **desulphurisation** unit using **S Zorb**™. Other ConocoPhillips refineries using S Zorb™ include those at Borger, TX, and Ferndale, WA.

World energy demand, growth in

The 2005 figure of 220 **million barrels per day of oil equivalent (MBDOE)** is just over twice the 1970 figure. Analysts predict that this will have risen to not less than 335 MBDOE by 2030.

World oil and gas consumption

In 2005, world consumption was 80 million **bbl** of oil and 300 billion ft^3 of gas. Below, these figures are normalised to **Million Barrels per Day of Oil Equivalent (MBDOE)**.

300 billion ft$^3 \equiv 8.1$ billion m^3

Now, 1 standard m^3 of any gas contains approximately 40 moles.

The heat of combustion of methane $= 889$ kJ mol^{-1}

Therefore, the heat released by the daily quantity of natural gas

$$= (8.1 \times 10^9 \times 40 \times 889 \times 10^3) \text{ J} = 2.9 \times 10^{17} \text{ J}$$

Since a **barrel** of oil is equivalent to 159 litre (0.159 m^3) and assigning a density of 950 kg m^{-3} and a **calorific value** of 43 MJ kg^{-1} to crude oil, the quantity of crude oil required to produce this energy on burning is

$$\frac{2.9 \times 10^{17}}{43 \times 10^6 \times 950} \times \frac{1000}{159} = 45 \times 10^6 \quad \textbf{bbl or 45 MBDOE}$$

The energy provided per day by oil and gas is:
$(80 + 45)$ MBDOE $= 125$ MBDOE

World energy consumption, according to figures given by ExxonMobil (the source of the oil and gas figures used above) is 220 MBDOE. The difference of 105 MBDOE, just over 45% of the total, is therefore provided by coal, nuclear fuels and renewables.

World oil refining capacity
The world oil refining capacity is estimated to be about 90 million **bbl** per day.

Wyoming, uranium in
Uranium was discovered in Wyoming in 1949, and a much larger find followed in 1953. Production of uranium as the basis of a nuclear fuel began in the 1950s and continued until 1992, during which 87.5 million kg of U_3O_8 were produced. Much site remediation was then required.

Uranium production in Wyoming was resumed in the year 2000 and the 2004 production was 600 tonnes. Other states of the US with significant uranium deposits are Arizona, Colorado, Texas, Florida, New Mexico and Utah. The Energy Bill signed by President George W Bush in August 2005 states that more nuclear power plants will be established in the US at the end of the current decade, and this obviously augurs well for domestic uranium production.

Wytch Farm, UK

Situated on the Dorset coast, this is the largest onshore oil and gas field in west Europe, currently producing 110 000 **bbl** per day. Recoverable reserves are about 300 million bbl. On its discovery in 1973 there was the difficulty that it is located at Poole, a seaside town of great natural beauty and environmental sensitivity. BP had been active in that part of England for some years by the time Wytch Farm was discovered and were the recipients of an award for innovation where the oil at Wytch Farm was accessed with negligible impact on the local area by a new technique called Extended Reach Drilling (ERD). This involved initial vertical drilling at an existing well outside Poole followed by horizontal drilling to extend the well to Wytch Farm. Later, drilling actually underneath Poole Bay took place in what became known as the 'Wytch Farm offshore extension'. **Liquefied petroleum gas** originating at Wytch farm is transferred to **Avonmouth**.

XYZ

Xijiang oil fields

Offshore fields in the South China Sea, about 80 miles from the Hong Kong coast, where operators including ConocoPhillips and Shell are active. ConocoPhillips have also recently been active in exploration and appraisal in the **Bohai Sea** and, using a **floating production, storage and offloading**, began production there in late 2002.

Yacheng gas field

The largest gas field in Chinese waters, discovered in 1983. It is in shallow water (about 90 metres). Gas from the field is carried 780 km by pipeline to Hong Kong, where it is used in electricity generation. There is also condensate, which is taken to nearby Hainan Island for storage and processing.

Yallourn

A town in the **Latrobe Valley** and the site of the first **brown coal** deposit to go into production, circa 1920. Coal winning is by removal of a shallow overburden and dredging of the coal at an open cut. The extent of the open cut expanded to such a degree that by the 1980s the township of Yallourn, previously having had a significant population, had to be sacrificed to it. Coal winning at Yallourn continues as it does at the other two major fields in the Latrobe Valley, namely Morwell and Loy Yang.

Yates field

Onshore oil field in Texas, discovered in 1926 and initially operated by the Ohio Oil Company, forerunner to Marathon Oil which until as recently as 2003 had an interest in the field. (Marathon Oil were sold to Kinder Morgan in 2003.) Initially pipelines belonging to the **Humble Oil and Refining Company** were, by a business arrangement, used to carry oil from the Yates field to storage vessels closer to centres of population where the oil could be marketed. By mid-1927 there were several wells at the field and about 9000 **bbl** of oil per day were being produced. In 1928, there was seepage of oil from one or more of the wells into the nearby Pecos River and thousands of bbl per day were recovered by skimming. From initially being a response to a contingency, skimming of oil which had migrated to the river became a routine operation and by 1933 over three million bbl of oil had been recovered in this way for subsequent storage and sale. Production peaked in 1929, the yield from the field for that year being 41 million bbl of crude oil. There was continuous operation over the subsequent decades and by 1985 the field had produced a billion bbl. **Enhanced oil recovery** techniques, including re-injection of **associated gas** and injection of seawater, have been used at Yates field since the 1970s. At the present time the field is producing about 7 million bbl annually and is about half way down the list of top one hundred US oil fields in terms of productivity.

Yellow tip

Effect sometimes observed either with a badly adjusted **burner** or when a substitute gas is used because the gas for which the burner was designed and adjusted is temporarily unavailable. The term is self-explanatory: the flame has a yellow tip. The effect is due to formation of carbon particles in the flame and is particularly undesirable in applications such as the melting of glass or steel where the carbon particles might enter the substance being melted and contaminate it. Even where contamination is not an issue, yellow tip signifies unsatisfactory burning because it affects the temperature profile within the flame.

YUKOS Oil Company

Post-Perestroika, privately-owned Russian Oil Company, which recently began exports of crude oil to the US. This situation is to some extent a repeat of history: in the early days of the oil industry, say in 1900, the US and Russia were about neck-and-neck in crude oil production. Other Russian oil companies include **Lukoil**, **Rosneft** and Transneft.

Zapolyarnoye

Major gas field in Russia operated by the Russian business group Gazprom. Proposals are in hand where Gazprom would trade part of Zapolyarnoye for a 25% interest in **Sakhalin Energy**.

Zeldovich mechanism

Also known as the Zeldovich reaction couple, it is the mechanism responsible for thermal **NOx**. Reaction takes place between oxygen radicals, nitrogen radicals and molecular nitrogen:

$$O + N_2 \leftrightarrow NO + N$$

$$N + O_2 \leftrightarrow NO + O$$

Zimbabwe, coal reserves in

There exist significant (> 400 Mt) reserves of **hard coal** in this country. Much of the coal produced annually is exported by rail to regional markets. Zimbabwe gets most of its electricity from the **Kariba dam** although there is some thermal generation, using local coal, at Hwange.

Zimbabwe, National Oil Company (NOCZIM)

This organisation co-ordinates imports of oil for Zimbabwe and, perhaps more importantly, determines pricing policy within the country for refined products including gasoline. The Zimbabwean economy, at the time of writing, is dismally bad and **NOCZIM** have been taking the desperate measure of making fuel at some of its outlets saleable for western currency only. Coupons can be bought for western currency and then exchanged for petrol.

Zirconia bricks

A **refractory material** for use in a **firebox**, distinguished from fireclay products by their much greater mechanical strength. Zirconia bricks have, however, a relatively poor resistance to slagging.

Further reading

Oil, Gas and Petrochemicals

Chemistry of petrochemical processes, Samir Matar and Lewis F Hatch, Butterworth-Heinemann 2001. ISBN 0-88415-315-0

Hydrocarbon process safety, J C Jones, Whittles Publishing 2003. ISBN 1-870325-54-0

Natural gas electric power in nontechnical language, Ann Chambers, PennWell 1999. ISBN 0-87814-761-6

Natural gas in nontechnical language, Rebecca L Busby (ed), PennWell 1999. ISBN 0-87814-738-1

Oil and natural gas: understanding the future, Pierre-René Bauquis and Emmanuelle Bauquis, Editions Hirle 2005. ISBN 2-914729-36-7

Oil refineries in the 21st Century, Ozren Ocic, Wiley-VCH 2005. ISBN 3-527-31194-7

Petrochemicals in nontechnical language, Donald L Burdick and William L Leffler, PennWell 2001. ISBN 0-87814-798-5

Petroleum catalysis in nontechnical language, John Magee and Geoffrey Dolbear, PennWell 1998. ISBN 6-87814-661-X

Petroleum chemistry and refining, James G Speight (ed), Taylor and Francis 1998. ISBN 1-56032-587-9

Petroleum production in nontechnical language, Forest Gray, PennWell 1995. ISBN 0-87814-450-1

Petroleum products: instability and incompatibility, George W Mushrush and James G Speight, Taylor and Francis 1995. ISBN 1-56032-297-7

Petroleum refining in nontechnical language, William L Leffler, PennWell 2000. ISBN 0-87814-776-4

Combustion, Incineration and Technology

Advanced pulverized coal injection technology and blast furnace operation, K Ishii (ed), Pergamon 2000. ISBN 0-08-043651-X

Advanced space propulsion systems, Martin Tajmar, Springer-Verlag 2003. ISBN 3-211-83862-7

Chemical history of a candle, Michael Faraday, reprint by Cherokee Publishing 1985. ISBN 0-87797-209-5

Coal combustion, Jerzy Tomeczek, Krieger Publishing 1994. ISBN 0-89464-651-6

Combustion and gasification of coal, A Williams, M Pourkashanian, J M Jones and N Skorupska, Taylor and Francis 2000. ISBN 1-56032-549-6

Combustion and incineration processes, Walter R Niessen, Marcel Dekker 2002. ISBN 0-8247-0629-3

Combustion engineering, Gary L Borman and Kenneth W Ragland, McGraw-Hill 1998. ISBN 0-07-115978-9

Elements of gas turbine propulsion, Jack D Mattingly, McGraw-Hill 1996. ISBN 0-07-114521-4

Energy efficiency for engineers and technologists, T D Eastop and D R Croft, Longman 1990. ISBN 0-582-03184-2

Energy from biomass: a review of combustion and gasification technologies, Peter Quaak, Harrie Knoef and Hubert Stassen, The World Bank, Washington DC. ISBN 0-8213-4335-1

Environmental chemistry: Asian lessons, Vladimir N Bashkin, Kluwer Academic 2003. ISBN 1-4020-1004-4

Fuels and combustion, Samir Sarkar, Orient Longman 1990. ISBN 81-250-0396-7

Flame and combustion, J F Griffiths and J A Barnard, Blackie Academic and Professional 1995. ISBN 0-7514-0199-4

Fuels and Engines, J C Guibet, Editions Technip 1999. Two volumes ISBN 2-7108-0751-3

Handbook of environmental engineering calculations, C C Lee and Shun Dar Lin, McGraw-Hill 1999. ISBN 0-07-038183-6

Incineration of hazardous, toxic and mixed wastes, James H Gill and John M Quiel, North American Manufacturing Co 1993. ISBN 0-9601596-4-9

Introduction to combustion: concepts and applications, Stephen R Turns, McGraw-Hill 2000. ISBN 0-07-116910-5

Introduction to energy: resources, technology and society, Edward Cassedy and Peter Grossman, Cambridge University Press 1998. ISBN 0-521-63767-8

Introduction to internal combustion engines, Richard Stone, Palgrave 1999. ISBN 0-333-74013-0

John Zink combustion handbook, Charles E Baukal (ed), CRC Press 2001. ISBN 0-8493-2337-1

Oxygen-enhanced combustion, Charles Baukal (ed), CRC Press 1998.
ISBN 0-8493-1695-2
Perry's chemical engineers' handbook, Robert H Perry and Don W
Green (eds), McGraw-Hill 1998. ISBN 0-07-115982-7
Power plant engineering, Black and Veatch, Chapman and Hall 1996.
ISBN 0-412-06401-4
Power plants: biofuels made simple, Brian Horne, Centre for Alternative
Technology Publications 1996. ISBN 1-898049-09-2
Principles of thermal sciences and their application to engineering, J C
Jones, Whittles Publishing 2000. ISBN 0-8493-0921-2
Rocket science, Alfred J Zaehringer and Steve Whitfield, Apogee Books
2004. ISBN 1-894959-09-4
Thermal solid waste utilisation in regular and industrial facilities, Lucjan
Pawlowski, Marzenna Dudzinska and Marjorie Gonzalez (eds),
Kluwer Academic 2000. ISBN 0-306-46449-7
Transport fuels technology, Eric M Goodger, Energy Institute 2000.
ISBN 0-9520186-2-4

Hydrogen and Fuel Cells

Fuel cell technology handbook, Gregor Hoogers (ed), CRC Press 2003.
ISBN 0-8493-0877-1
*Powering the future: the Ballard fuel cell and the race to change the
world*, Tom Koppel, John Wiley & Sons 1999. ISBN 0-471-64421-8
*Tomorrow's energy: hydrogen, fuel cells and the prospects for a cleaner
planet*, Peter Hoffmann, MIT Press 2002. ISBN 0-262-58221-X

Renewable Energy

Carbon and its domestication, A M Mannion, Springer 2006. ISBN 1-
4020-3957-3
Cool energy: renewable solutions to environmental problems, Michael
Brower, MIT Press 1998. ISBN 0-262-52175-0
*Energy options: an introduction to small scale renewable energy technol-
ogy*, Drummond Hislop (ed), Intermediate Technology Publications
1992. ISBN 1-85339-082-8
Energy systems and sustainability: power for a sustainable future,
Godfrey Boyle, Bob Everett and Janet Ramage (eds), Oxford/The
Open University 2003. ISBN 0-19-926179-2
Green energy: biomass fuels and the environment, United Nations
Environment Programme, UNEP 1991. ISBN 92-807-1308-6
Renewable energy in nontechnical language, Ann Chambers, PennWell
2004. ISBN 1-59370-005-9
Renewable energy: power for a sustainable future, Godfrey Boyle (ed),
Oxford/The Open University 2004. ISBN 0-19-926178-4

Useful Units and Conversions

Unit	Symbol	SI units
Joule	J	$kg \cdot m^2/s^2$
Calorie	Cal	4.19 J
British Thermal Unit	BTU	1.054 kJ
Cubic foot of natural gas	ft³ natural gas	1,000 BTU
		1.055 MJ
Therm	Therm	100,000 BTU
		105.5 MJ
ton of TNT	tTNT	1 Gcal
		4.184 GJ
Barrel of oil equivalent	bboe	5.8 MBTU
		6.12 GJ
ton of oil equivalent	TOE	10 Gcal
		41.868 GJ
Watt	W	$kg \cdot m^2/s^3$
BTU per hour	BTU/h	0.293 W
Calorie per second	cal/s	4.1868 W
Horsepower	hp	736 W
Kelvin	K	$T[K] = T[°C] + 273.15$
Degree Celsius	°C	$T[°C] = T[K] - 273.15$
Degree Rankine	°R or °Ra	$T[°R] = 1.8 \times T[K]$
Degree Fahrenheit	°F	$T[°F] = (T[K] \times 1.8) - 459.67$
		$T[°F] = (T[°C] \times 1.8) + 32$
Net Calorific Value[*]	MJ/kg	Gross CV – 0.212 H – 0.024 M
Net Calorific Value[*]	kcal/kg	Gross CV – 50.7 H – 5.83 M
Net Calorific Value[*]	BTU/lb	Gross CV – 91.2 H – 10.5 M
kilocalorie per kilogram	kcal/kg	238.8 MJ/kg
BTU per pound	BTU/lb	1.800 kcal/kg
		429.9 MJ/kg

[*] H = hydrogen content, M = moisture content

Useful units and conversions

	Multiples				**Submultiples**	
E	exa	$\times 10^{18}$		a	atto	$\times 10^{-18}$
P	peta	$\times 10^{15}$		f	femto	$\times 10^{-15}$
T	tera	$\times 10^{12}$		p	pico	$\times 10^{-12}$
G	giga	$\times 10^{9}$		n	nano	$\times 10^{-9}$
M	mega	$\times 10^{6}$		μ	micro	$\times 10^{-6}$
k	kilo	$\times 10^{3}$		m	milli	$\times 10^{-3}$
h	hecto	$\times 10^{2}$		c	centi	$\times 10^{-2}$
da	deca	$\times 10$		d	deci	$\times 10^{-1}$

Fuel and energy terms in other languages

Barrel
albermil: Arabic
bareru: Japanese
barilotto: Italian
barril: Spanish, French
boshkeh: Persian
das Faß: German
koyala: Hindi

Bitumen
betún: Spanish

Bituminous coal
steenkool: Dutch, Afrikaans
die Steinkohle: German

Briquette
das Brikett: German
das Steinkohlenbrikett: (German word for briquette made from bituminous coal)

Brown coal
die Braunkohle: German

Butane
butan: Japanese

Coal (without any reference to rank)
amalahle: Xhosa and Zulu (Bantu languages, spoken by indigenous South Africans)
batu arangand: Malay
batubara: Indonesian
hiili: Finnish

kömür: Turkish
sekitan: Japanese
taan hin: Thai
uhel: Czech

Coke
ko-ke: Japanese
koka: Hindi

Crude oil
das Erdöl: German (also, *das Rohöl*)
gen'yu: Japanese
kacca tela: Hindi
yeryaði: Turkish

Diesel
gasolio per autotrazione: Italian
gazoleo: Spanish
keiyu: Japanese

Electricity
sähkö: Finnish

Fuel (in the broad sense)
polttoaine: Finnish

Gasification
vergasung: German

Gasoline
bensin: Norwegian
chair yow: Cantonese
minyak patrol: Malay
tian yew: Mandarin

Fuel and energy terms in other languages

Hydroelectric installation
wasserkraftwerk: German

Kerosene
das Flugbenzin: German
for sway: Cantonese
gazyaði: Turkish
huo yew: Mandarin
kewozin: Creole (used in parts of Caribbean)
minyak tanah: Malay
naum maun gas: Thai
parafin: Norwegian
petrol lampant: Romanian
toh-yu: Japanese
zayt al-barafeen: Arabic (also *al-kayruseen* and *zayt al-kaaz*)

Lead-free gasoline
bleifrei: German
loodvrij: Dutch
sans plomb: Italian
sin plomo: Spanish

Lignite
lìnyìt: Turkish

Natural gas
das Erdgas: German

Nuclear power
nâbhikîya úakti: Hindi

Oil
alnapht: Arabic
huil: French
luil: Creole (used in parts of Caribbean)
mafuta: Swahili (Bantu language)
napht: Persian
oli: Xhosa (dominant in Transvaal)
olie: Norwegian, Danish
olje: Dutch, Afrikaans
oyili: Zulu (dominant in Kwazulu-Natal)
sekiyu: Japanese (also, *abura* and *oira*)

ulei: Romanian
zet: Lebanese
öljy: Finnish

Oil tanker
der Oltanker: German (sometimes taken to mean supertanker)
petrolier: Romanian
der Tanker: German

Peat
der Torf: German

Propane
puropan: Japanese

Pulverised fuel
der Kohlenstaub: German

Refinery
seirensho: Japanese

Retort coal gas
das Leuchtgas: German

Sawdust
sågspån: Swedish

Shale oil
schieferöl: German
she-ruoiru: Japanese

Steam turbine
die Dampfturbine: German

Terminal
der Olhafen: German

Uranium
uraniumu: Japanese
yurçniyama: Hindi

Wood
kayu-kayan: Malay